SPECTROSCOPIC TECHNIQUES

and

HINDERED MOLECULAR MOTION

SPECTROSCOPIC TECHNIQUES
and
HINDERED MOLECULAR MOTION

Ferid Bashirov

CRC Press
Taylor & Francis Group
Boca Raton London New York

CRC Press is an imprint of the
Taylor & Francis Group, an **informa** business

CRC Press
Taylor & Francis Group
6000 Broken Sound Parkway NW, Suite 300
Boca Raton, FL 33487-2742

First issued in paperback 2019

ISBN-13: 978-1-4398-7083-9 (hbk)
ISBN-13: 978-0-367-38228-5 (pbk)

Library of Congress Cataloging-in-Publication Data

Bashirov, Ferid.
Spectroscopic techniques and hindered molecular motion / Ferid Bashirov.
p. cm.
Includes bibliographical references and index.
ISBN 978-1-4398-7083-9 (hardback)
1. Molecular structure. 2. Molecular rotation. 3. Molecular crystals. 4. Spin-lattice relaxation. 5. Molecular dynamics. I. Title.

QD461.B267 2012
541'22--dc23 2011020080

Visit the Taylor & Francis Web site at
http://www.taylorandfrancis.com

and the CRC Press Web site at
http://www.crcpress.com

Contents

Preface

This monograph presents an advanced theoretical approach to the description of the spectroscopic study of the classical local thermal motion of small molecules and molecular fragments in crystals. The so-called extended angular jump model, which explores neither hydrodynamic theory nor dynamical methods, approximates the motion. This model of hindered molecular motion (HMM) is intermediate between the rotational diffusion model and the model of fixed angular jumps. Accounting for two point-symmetry groups of rotation—the symmetry group of molecular motion and that of its site, including distortion—enhances the power of the model. Symmetrized upon the irreducible representations of the crystallographic point symmetry group, the hindered states of the motion handle the transformation property of molecular quantities. Two phenomenological parameters of the model—the symmetrized correlation times and the dynamical weights of the motion—explore the fitted dynamic quantities of the theory. They allow us to obtain expanded knowledge of molecular dynamics as well as crystalline structure. We applied this innovative HMM theory to derive new analytical expressions describing quantitatively nuclear magnetic resonance relaxation rates and homogeneously broadened spectral lines offered by dielectric, infrared, Raman, Rayleigh, and incoherent neutron scattering spectroscopy techniques for both mono- and polycrystalline samples. The presented experimental spectroscopic investigations of HMM in molecular condensed substances make obvious the effectiveness of simulating HMM in the framework of the extended angular jump model. For example, we demonstrate that the experimental anisotropy of proton spin-lattice relaxation in ammonium chloride observed at low temperatures originates by structural distortion of the crystalline cubic unit cell. In addition, we establish theoretically that the nature of elastic and quasi-elastic incoherent neutron scattering is directly connected with the symmetry properties of HMM.

The treatment is classically phenomenological and should be useful primarily to experimentalists, and so we have included detailed derivations, tables of mathematical and numerical results, and comparisons of the theory with the experiments. Scientists, lecturers, and students of condensed matter physics can use the proposed monograph as a textbook as well as a manual in order to study microscopic peculiarities of the dynamics and the structure of molecular substances. Basic knowledge of the theory of stochastic processes, the theory of representations of the point symmetry groups, and some experimental practice in spectroscopy techniques would be beneficial in order to fruitfully exploit the contents of the book. I have not attempted to give a complete list of references, but I have occasionally inserted them when lack of space prevented me from making more than a mere allusion to the contents of the paper quoted.

Introductory comments added to some chapters will help graduate students readily assimilate the materials. I believe the book will stimulate new ideas and allow readers to better understand the structure and dynamics of condensed molecular media in the years to come. I would appreciate any comments as to possible errors or concerning corrections and improvements. At the same time, I expect that the statements presented here will find friendly acceptance and will be content if this book helps transfer some of the delight that I have experienced during my research in this fascinating field.

Acknowledgments

I would like to acknowledge the collaboration of my colleague Professor Nail Gaisin (Kazan Technological University). I take this opportunity to express my sincerest gratitude to Professor Albert Aganov (Kazan Federal University) for his generous and sympathetic help. Finally, I thank my wife, Khadycha, for her patience and encouragement; she made many efforts in order to give me the time I needed to write this book.

Introduction

Determining the nature of intra- and intermolecular interactions as well as the character of molecular motion is one of the key problems of condensed state physics. Intermolecular interactions are feeble in gases and the motion of molecules is free during a mean time between two sequential collisions. The strong intermolecular interaction in a condensed substance converts a problem of statistical description of molecular motion into a many-body problem. Difficulties exhibited by dynamical and statistical theories of continuous medium have led to the involvement of alternative research techniques based on hypothetical models of the thermal molecular motion. At the same time, any model for approximating a physical phenomenon requires rigorous checking for its correspondence to actual events. Such checking is to be performed by experiments on real samples or, lacking those, by means of computer modeling.

Conformity of theoretical processing to experimental data and consistency of the results obtained by various experimental procedures are the major factors of quality in physical theory. Difficulties arising from the quantitative description of spectroscopic data that provide condensed molecular media give evidence to the fact that microscopic-level mysteries persist even today. Considering that the common physical properties of molecular solids and liquids are similar at temperatures close to fusion point, the solution of the thermal molecular motion problem in solids could accelerate elaboration of the general theory of condensed molecular media. This lack of a general theory impedes an understanding of the collected experimental data and the setup of new research.

The solution of the molecular motion problem in crystals has further value in the applied sense. A list that is by no means complete of those processes, phenomena, and understandings that would benefit from the study of molecular motion in crystals would include an understanding of the nature of structural phase transformation; the process of premelt crystallization; the phenomena of piezoelectricity, ferroelectricity, double refraction, and optical rotation; and the determination of the geometry of potential crystalline field, interatomic distances, the probabilities of elementary rotational displacements, and the energy barriers of molecular motion.

Liquid crystals present a classroom of substances for which studying orientational ordering shows a special significance. Organic compounds, polymers, pharmaceutical products, and biological systems consist of molecular fragments that possess rotational or conformational degrees of freedom, or there is an atomic exchange within. Determining laws of molecular motion allows one to obtain knowledge of that process and in what degree the interactions of various natures depend on each other. For example, magnetic and electrical interactions of nuclei and electrons depend on mechanical interactions and vice versa, as well as on external conditions: temperature, pressure, electrical and magnetic fields; composition and quantity of impurities; structural distortion, and so on. And knowledge of these laws stimulates growing new crystals and synthesizing new substances with the desired properties. It is clear that the noted examples do not exhaust all the various scientific and engineering problems for which experimental and theoretical study of the thermal molecular motion problem in condensed substances has paramount value.

The scope of this monograph is restricted to small molecules and molecular fragments such as N_2, $HC1$, CO_2, CH_4, H_2O, NH_4, BeF_4, NH_3, CH_3, C_6H_6, SF_6, and other symmetrical atomic formations that exhibit local hindered motion in molecular condensed media: molecular and ionic crystals, molecular liquids, liquid crystals, polymeric solids, and biological objects. Chapter 1 presents the actual state of the theory of hindered molecular motion (HMM). Peculiarities of two well-known HMM models—the rotational diffusion model (RD model) and the model of fixed angular jumps (FAJ model)—are discussed. In Chapter 2, we describe the extended angular jump model (EAJ model) resulting from a selective choice of the priority ideas underlying two former models; then, by using the EAJ model, a rigorous solution of the HMM stochastic problem for molecular crystals is given.

The general and explicit forms of the angular autocorrelation functions of the unitary spherical tensor components adapted to the crystallographic point symmetry groups are derived for the EAJ model in Chapter 3.

The molecular spectroscopy applications of the HMM theory are given in Chapters 4–6. Frequency-dependent line shapes in dielectric, infrared absorption, and Rayleigh and Raman light scattering spectra show like peculiarities (Chapter 4). They consist of the sum of Lorentzians whose quantity is determined by the number of irreducible nonequivalent, nonidentical representations of a HMM symmetry group and do not exceed the number 3.

The application of the HMM theory to discussion of the exponential nuclear magnetic resonance relaxation is given in Chapter 5. We are concerned with the spin-lattice relaxation of the dipole and quadruple nuclei with respect to the laboratory and rotating reference frames in single- and polycrystalline samples. Explicit expressions are given for the relaxation rates in three- and four-spin systems. The experimental anisotropy of the proton relaxation rates in cubic ammonium chloride is explained by the tetragonal distortion of its cubic unit cell at low temperatures. A decrease of the proton relaxation rates in rotating three-spin systems with respect to predictions following the Bloembergen-Purcell-Pound (BPP) relaxation theory is interpreted in the framework of the EAJ model.

Incoherent neutron scattering application is discussed in Chapter 6. It is shown that HMM symmetrized solely in identical, irreducible representations of the point groups contributes to elastic incoherent neutron scattering; nonidentical representations originate quasi-elastic scattering. The interpretation of the experimental data on elastic incoherent neutron scattering in some polycrystalline samples is revised, and the realistic knowledge of the hindered motion of proton vectors and their site symmetry are obtained.

About the author

Ferid Israphilovich Bashirov is currently a professor and chair of the Teaching Laboratory of Electricity and Magnetism in the General Physics Department, Institute of Physics, Kazan Federal (State) University, Kazan, in the Russian Federation. Here, he studied physics and received his PhD in the physics of magnetic phenomena in 1972 and his DSc in the physics of condensed matter in 2006. Dr. Bashirov was a lecturer for general physics at the University of Oran (Algeria) from 1975 to 1978 and at the University of Conakry (Guinea) from 1998 to 2010.

Dr. Bashirov's research interests include the experimental and theoretical study of the dynamics and structure of condensed molecular media by spectroscopic techniques. His scientific and technical contributions include manufacturing the coherent pulsed nuclear magnetic resonance (NMR) high-performance spectrometer for laboratory purposes; the growth of crystals with internal molecular motion from aqueous solutions; the discovery of the existence of the tetragonal structure distortion in the ordered phase of cubic ammonium chloride by studying the anisotropic proton magnetic spin-lattice relaxation therein; the elaboration of the extended angular jump model for hindered molecular motion in crystals; and the development of the theory of spectroscopic technique applications to the study of hindered molecular motion in mono- and polycrystalline substances.

He has taken part in several international scientific meetings: Congress AMPERE–1994 (Russia), ICORS–1996 (United States), Nuclear and Electron Relaxation Workshop–1997 (Italy), Congress AMPERE–1998 (Germany), Colloque AMPERE–1999 (Lithuania), Conference QENS–2004 (France), Conference PLM MP–2005 (Ukraine), and others. His scientific and teaching publications consist of 105 papers, articles, manuscripts, and books.

Dr. Bashirov has been recognized by International Biographical Centre (Cambridge, England) as a member of "Top 100 Scientists 2011" and included in 28th Edition of the 2011 Index *Who's Who in the World.*

chapter one

Fundamentals of the theory of hindered molecular motion

1.1 *Problem of hindered molecular motion*

Development of the theory binding the structure of a substance to its physical properties is a significant trend in modern study of condensed matter physics. This theory allows one to process the physical, chemical, and technological properties of a chain of like substances without synthesizing them, using only structural data and a minimal number of experimental parameters obtained for the few samples of a chain.

In crystals, owing to the predominance of attractive interatomic forces over repulsive forces, atoms and molecules have an ordered disposition with periodic local points and spatial structures. In molecular crystals, nodes of a crystal lattice are occupied by molecules, and coupling between them is carried out by weak Van der Waals's forces, mainly dispersion ones. Potential energy of an intermolecular bond is on the order of 10 kJ per mole. In such crystals, rotational mobility of molecules—and hence an uncertainty in the angular positions of molecules—is probable [1–3].

Fewer examples of molecular crystals are presented by a large ensemble of inorganic and organic compounds: ammonia NH_3, dry ice CO_2, solid hydrogen sulfide H_2S, normal alkenes C_nH_{2n+2}, benzyl C_6H_6, amino acids $CHNH_3^+COO$–R, adamantane $C_{10}H_{16}$, nitrate(V) of tetraaquacalcium $[C_a(H_2O)_4](NO_3)_2$, perchlorate of tetraaquacalcium $[C_a(H_2O)_4](ClO_4)_2$, perchlorate of tetraaminozinc $[Z_n(NH_3)_4](ClO_4)_2$, triglycine sulfate $(CH_2NH_2COOH)_3H_2SO_4$, pyrrolidinium hexachloroantimonate(V) $[C_4H_8NH_2][SbCl_6]$, sodium ammonium selenate dihidrate $NaNH_4SeO_4{\cdot}2H_2O$, trisguanidinium heptafluorozirconate $[C(NH_2)_3]_3ZrF_7$, and others. Molecules NH_3, CO_2, H_2S, CH_4, C_6H_6, and $C_{10}H_{16}$, and atomic groups CH_2, CH_3, H_2O, and NH_3, considered classical solid bodies, occupy orientation potential wells separated by finite energy barriers and can execute rotational jumps between these wells. The mobile polyatomic ions, such as ions of ammonium NH_4^+, fluoroborate BF_4^- or fluoroberillate BeF_4^{2-}, are part of ionic crystals—NH_4Cl, NH_4F, NH_4Br, NH_4I, $(NH_4)_2SO_4$, $NH_4H_2PO_4$, $(NH_4)_2HPO_4$, $(NH_4)_2BeF_4$, NH_4BF_4, and so forth—and are also orientationally disordered. Water molecules are mobile in crystal hydrates $CuSO_4*5H_2O$, $CuCl_2*2H_2O$, $K_2CuCl_4*2H_2O$, sodium ammonium selenate dihidrate $NaNH_4SeO_4*2H_2O$, and so on.

A random reorientation of molecules or their fragments in crystals associated with classical thermal process is called hindered molecular rotation [4]. In the general case of thermoactivated, classically arranged molecules originated by molecular rotation, atomic exchange, or conformation transition, we shall call such motion *hindered molecular motion* (HMM) [5]. It occurs at not very low temperatures ($T \geq 40$ K). HMM affects the physical properties of crystals. As a rule, molecular crystals show a low melting point (up to 500 K). Several structural modifications are often observed therein. Peculiarities of HMM depend on the composition and structure of molecules, their interaction with nearest neighbors, the thermodynamic state of the substance, and the external conditions. Owing to microscopic sizes of molecules, direct methods for investigating HMM do not exist.

Fortunately, there are many indirect experimental procedures for studying the physical properties of matter at microscopic levels. The most fruitful methods exploit spectroscopy techniques such as dielectric spectroscopy [6–9], infrared spectroscopy [10–14], molecular light scattering spectroscopy [11–15], spectroscopy of nuclear magnetic resonance (NMR) relaxation [16–19], and spectroscopy of incoherent neutrons scattering [20–24]. By using a suitable theory, we can determine moving atoms and structure parameters of a moving system as well as frequency range, activation energy, kind of motion, and the effect of phase transition to motion.

Dynamic parameters of HMM are enclosed in relaxation constants and the shape of spectral lines. The modern theoretical procedure that describes relaxation rates and line shapes deals with calculating the autocorrelation function (ACF) of the random molecular physical quantities. Results of ACF processing depend on both the peculiarities of the molecular physical quantities and the hypothetical model of molecular motion [6–34].

Extensive experimental and theoretical knowledge of local molecular motion in condensed substances is being collected today. Studying the local motion of molecules is reliable in liquids. In solid substances, however, the treatment of thermal motion of molecules is ambiguous. In spite of the fact molecular crystals and liquids are more alike than distinct in basic physical properties, a consistent description of the local HMM is absent.

Dominating HMM theories were developed within the frameworks of two diverse HMM physical models: the rotational diffusion model (RD model) [6] and the fixed angular jump model (FAJ model) [16]. However, the anisotropic properties of classical local HMM were not taken explicitly into consideration, and the available theories did not yield a guarantee of quantitative agreement between the theoretical and experimental data.

The effect of a crystalline environment on HMM was not investigated, and the theory of solid state HMM was not adjustable to discuss the liquid state local HMM.

Recently, we suggested an advanced physical model termed the *extended angular jump model* (EAJ model), where we found an approach to the general solution of the HMM problem [35, 36]. It concerns symmetrical molecules that exhibit a classical random motion in the local crystalline surroundings. This book intends first to present this theory in all details (Chapters 1–3) and then to give its application to the quantitative discussion of the experimental data derived by using common molecular spectroscopy techniques (Chapters 4–6).

1.2 Basis of the angular autocorrelation function technique

In a system of uncorrelated molecules, the time-dependent autocorrelation function (ACF) $K(t_1,t_2)$ of a random classical molecular variable $F(t)$ is the statistical average of the product of the values of this variable determined at two different instants, t_1 and t_2:

$$K(t_1,t_2) = \left\langle F(t_1) \cdot F(t_2)^* \right\rangle \tag{1.1}$$

where the asterisk denotes complex conjugation. In our treatment, we consider macroscopic thermodynamic properties of the molecular system to be independent of time, a simple stationary Markovian chain describes the stochastic process, and the ergodic theorem is valid. Therefore, the ACF does not depend at once on both fixed times t_1 and t_2. It depends only on t, the time interval between the two events: $t = t_2 - t_1$. Moreover, the instant t_1 (or t_2) can be chosen arbitrarily, as well as equal to zero. Consequently, we can rewrite Equation (1.1) as

$$K(t_1,t_2) = K(t_2 - t_1) = K(t) = \langle F(0) \cdot F(t)^* \rangle \tag{1.2}$$

Usually, the variable $F(t)$ is an explicit function of the molecular radius vector $\mathbf{b}$, which depends implicitly on time t:

$$F(t) = F(\mathbf{b}) = F[\mathbf{b}(t)] \tag{1.3}$$

For a continuous function $F(\mathbf{b})$, the ACF can be derived by the following double-space integration [37]:

$$K[\mathbf{b}(t)] = \oiint F(\mathbf{b}_0) F(\mathbf{b})^* \, W(\mathbf{b}_0, t, \mathbf{b}) \, \mathrm{d}\,\mathbf{b}_0 \, \mathrm{d}\,\mathbf{b} \tag{1.4}$$

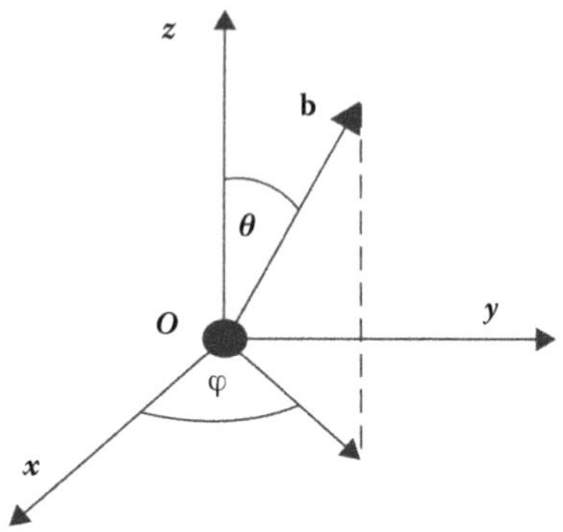

Figure 1.1 Polar angles of the radius vector $\mathbf{b} = \mathbf{b}\,(b, \theta, \varphi)$.

where $W(\mathbf{b}_0, t, \mathbf{b})$ is the time-dependent binary distribution function of the vectors $\mathbf{b}_0$ and $\mathbf{b}$, where $\mathbf{b}_0$ is the vector $\mathbf{b}$ determined at the initial instant of time $t = 0$.

In order to unite the treatment of different spectroscopic applications, we expand the quantity $F(\mathbf{b})$ in series of the spherical harmonics $Y_m^{(\nu)}(g)$, the components of the unitary spherical tensor $Y^{(\nu)}(g)$ of the rank ν:

$$F(\mathbf{b}) = \sum_{\nu} F^{(\nu)}(\mathbf{b}) = \sum_{\nu, m} F_m^{(\nu)}(\mathbf{b}) = \sum_{\nu, m} F_m^{(\nu)}(b, g) = \sum_{\nu, m} f_m^{(\nu)}(b)\, Y_m^{(\nu)}(g) \qquad (1.5)$$

where b is the modulus of $\mathbf{b}$ and g is its space orientation angle determined in the polar reference frame $g = (\theta, \varphi)$, θ is the polar angle, and φ is the azimuth angle (Figure 1.1). The orthonormalization condition of the unitary spherical tensor components is

$$\oint Y_m^{(\nu)}(g)^*\, Y_{m'}^{(\nu')}(g)\, dg = \delta_{\nu\nu'}\delta_{mm'} \qquad (1.6)$$

where $dg = \sin\theta\, d\theta\, d\varphi$ is the infinitely small element of solid angle, and $\delta_{\nu\nu'}$, $\delta_{mm'}$ are Kronecker deltas. We would like to note that θ takes values from 0 up to 180 degrees ($0 \le \theta \le \pi$ rad), the limits of φ are $0 \le \varphi \le 2\pi$ rad.

The functions $Y_m^{(\nu)}(g)$ are the basis functions of the irreducible representations of the continuous group of three-dimensional rotations O^{3+}. They transform each other as [38, 39]:

$$Y_m^{(\nu)}(g) = \sum_{n=-\nu}^{\nu} D_{mn}^{(\nu)}(\Omega)\, Y_n^{(\nu)}(g') \qquad (1.7)$$

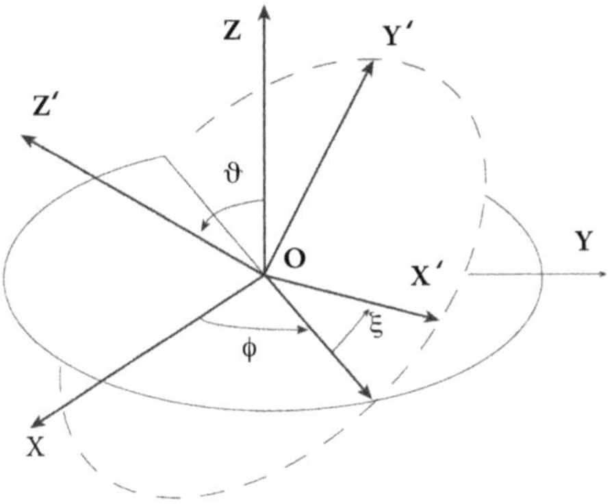

Figure 1.2 Euler angles between two rectangular reference frames: ϕ, the azimuthal angle; ϑ, the polar angle; and ξ, the angle of pure rotation.

where $D_{mn}^{(\nu)}(\Omega)$ is an element of Wigner transformation matrix (an element of the matrix of irreducible representations $D^{(\nu)}$ of the group O^{3+}, Wigner function), $\Omega \equiv (\varphi, \vartheta, \xi)$ is the three-dimensional Euler angle between two orthogonal reference frames, in which the functions $Y_m^{(\nu)}(g)$ and $Y_n^{(\nu)}(g')$ are determined (Figure 1.2).

Taking into consideration that the translational motion of molecules is frozen in solids, all $f_m^{(\nu)}(b)$ must be taken as constant factors of b in Equation (1.5), and we can present Equation (1.4) in the form

$$K[\mathbf{b}(t)] = \sum_{\nu,m} \sum_{\nu',m'} f_m^{(\nu)}(b) f_{m'}^{(\nu')}(b)^* \oiint Y_m^{(\nu)}(g_0) Y_{m'}^{(\nu')}(g)^* \, W(g_0,t,g) \mathrm{d}g_0 \mathrm{d}g \quad (1.8)$$

which is the general design formula of the ACF. For a variable $F(\mathbf{b})$ presented by the single spherical tensor component $F_m^{(\nu)}(\mathbf{b}) = f_m^{(\nu)}(b) Y_m^{(\nu)}(g)$, Equation (1.8) reduces to

$$K_m^{(\nu)}[\mathbf{b}(t)] = \left| f_m^{(\nu)}(b) \right|^2 \oiint Y_m^{(\nu)}(g_0) Y_m^{(\nu)}(g)^* \, W(g_0,t,g) \mathrm{d}g_0 \mathrm{d}g \quad (1.9)$$

In accordance with the existence condition, $W(g_0,0,g)$ obeys the normalization conditions:

$$\oint W(g_0, 0, g) \mathrm{d}g = 1/4\pi \quad (1.10)$$

$$\oiint W(g_0, 0, g) \, \mathrm{d}g \mathrm{d}g_0 = 1 \quad (1.11)$$

The initial value of probability density $W(g_0,0,g)$ has a uniform distribution of the angle in the framework of the RD model as well as in cases when the molecular neighborhood is isotropic:

$$W(g_0,0,g)=\frac{1}{4\pi}\delta(g-g_0) \tag{1.12}$$

where $\delta(g-g_0)$ is Dirac delta-function.

At last, in order to unite the different spectroscopic applications, we shall normalize (factorize) the ACF by

$$K_m^{(\mathrm{v})}(t)=\frac{K_m^{(\mathrm{v})}[\mathbf{b}(t)]}{\left|f_m^{(\mathrm{v})}(b)\right|^2}=\oiint Y_m^{(\mathrm{v})}(g_0)Y_m^{(\mathrm{v})}(g)^*\,W(g_0,t,g)\,\mathrm{d}g_0\,\mathrm{d}g \tag{1.13}$$

Using Equation (1.13) allows us to calculate the normalized ACF $K_m^{(\mathrm{v})}(t)$ of the unitary spherical tensor component $Y_m^{(\mathrm{v})}(g)$, which has no dependence of the physical quantity. Meanwhile, it retains the stochastic properties of hindered molecular motion. The processing of this ACF requests knowledge of the explicit form $W(g_0,t,g)$, the time-depended binary distribution function of the molecular vector orientation, which links simply to the HMM model.

1.3 Autocorrelation functions adapted to the RD and FAJ models

A systematic theoretical study of local HMM was initiated by Debye, who has solved the problem of the rotational diffusion motion of a rigid molecule identified with reorientation of a Brownian particle in a viscous medium [6]. According to this model, the molecular rotational displacement to a finite angle (of about 1 steradian) takes place by means of an infinite number of angular displacements to infinitely small angles random in direction and time. Since then, the thermal rotations of "large" molecules in liquids were approximated successfully by using the model of Debye or the simple rotational diffusion model (SRD model). Valiev advanced the SRD model by using the representation theory of continuous rotation groups [40].

Further progress in HMM theory was achieved by solving the "rotational random walk problem" of a molecular top in the isotropic matrix [29, 41]. The equation of Chapman and Kolmogorov in the integral form has been applied to solve the stochastic problem for the jump rotational diffusion model (JRD model). The uniform distribution relates to the instantaneous angular orientation of a molecule, and its random local motion takes place by means of rotational displacements to the finite angles, the values of which have a continuous bell-shaped distribution.

The ACFs derived in the frameworks of the SRD and JRD models differ from each other by the explicit expressions of the sole dynamical parameter $\tau^{(\nu)}$, the correlation time. For instance, in the case of hindered rotation of spherical molecules, the ACF expression is

$$K_m^{(\nu)}(t) = \frac{1}{4\pi}\exp\left(-\frac{t}{\tau^{(\nu)}}\right) \tag{1.14}$$

In the framework of the SRD model, the correlation time $\tau^{(\nu)}$ depends on the constant of isotropic rotational diffusion D and the rank ν of the molecular physical quantity tensor

$$\tau^{(\nu)} = \frac{1}{\nu(\nu+1)D} \tag{1.15}$$

In the framework of the JRD model, $\tau^{(\nu)}$ depends on the average time between two successive steps of rotational displacement τ, the probability density $p(\Omega)$ of the unitary rotation to the angle Ω, and the rank ν of the molecular physical quantity tensor

$$\tau^{(\nu)} = \frac{\tau}{1 - \oint p(\Omega)\, \boldsymbol{T}^{(\nu)}(\Omega)\,\mathrm{d}\Omega} \tag{1.16}$$

where $\boldsymbol{T}^{(\nu)}(\Omega)$ is the operator for rotational displacement.

The macroscopic feature of HMM based on the hydrodynamic theory of viscosity underlies the SRD model, whereas the microscopic statistical behavior of molecules is the basis of the JRD model. There is experimental evidence that the correlation times of rotational diffusion of large molecules in liquids taken from the spectra of NMR relaxation and light scattering are not equal to those taken from the dielectric and infrared spectroscopy. The dependence of the correlation times of the rank of the molecular quantity tensor is consistent with the frameworks of both rotational diffusion models. However, both of them came into disagreement by simulating the HMM of small molecules, whose correlation times do not depend on the rank of the molecular quantity tensor [42–44].

Moreover, according to Neumann's principle of crystal physics, any molecular physical quantity, as well as the correlation function, would reflect the anisotropic properties of solids [1–3, 45]. Meanwhile, the angular ACFs derived in the frameworks of the SRD and JRD models are always isotropic for the molecular tops, which display two and three rotational degrees of freedom. Such divergence from Neumann's principle takes place in spite of the fact that the concept of anisotropy of the molecular motion persists in these theories, for example, by means of such outcomes, as the tensor of anisotropic rotational diffusion and the probability density of anisotropic reorientation. The isotropy of correlation functions lets us discern two diffusion models. In further consideration, we shall call both of them the rotational diffusion model (RD model).

Bloembergen introduced a discrete HMM model, the model of fixed angular jumps (FAJ model), in the framework of which the rigid molecule occupies only a limited set of the angular positions and exhibits a classical rotation between them by means of random jumps to the finite angles [16]. The probability equation in finite differences was applied to solve the stochastic problem [16, 28, 46–48]. We present an advanced implicit expression of the ACF derived in the FAJ model framework by

$$K_m^{(\nu)}(t) = \sum_{j,\alpha} \left[h_{\alpha m}^{(\nu)}(\theta_j, \varphi_j) \right] \exp\left(-\frac{t}{\tau_\alpha} \right) \tag{1.17}$$

where the factors $h_{\alpha m}^{(\nu)}(\theta_j, \varphi_j)$ are analytical functions symmetrized on the irreducible representations of the point symmetry group of rotation. Here, θ_j and φ_j are the values of polar and azimuthal angles of a molecular vector $\mathbf{b} \equiv \mathbf{b}(b, \theta_j, \varphi_j)$ fixed in the laboratory reference frame, j enumerates the equilibrium directions of the vector $\mathbf{b}$, and α labels the irreducible representations.

With respect to Equation (1.17), the correlation function consists of the sum of exponentially decreasing terms, for which quantity is equal to the number of nonidentical, irreducible representations of the HMM symmetry group. The correlation times τ_α depend on the average time between two consequent angular jumps τ, the probabilities of class rotational displacement p_i, the class characters $\chi_{i\alpha}$, and the dimensions μ_α of the irreducible representations Γ_α by

$$\tau_\alpha = \frac{\tau}{1 - \frac{1}{\mu_\alpha} \Sigma_i p_i \chi_{i\alpha}} \tag{1.18}$$

It should be noted that the FAJ model, as well as the JRD model, are not linked to the hydrodynamic characteristics of the medium. This is an advantage of both models. The FAJ model found satisfactory application by discussing the nuclear magnetic resonance relaxation experiments in some powder samples with internal HMM [16, 46–48]. However, in molecular single crystals, the experimental NMR relaxation rates did not agreed with those calculated theoretically by using the FAJ model [49–51].

chapter two

Solution of the stochastic problem for the extended angular jump model

2.1 *Description of the extended angular jump model*

Solving a physical problem by accounting for the symmetry properties of the substance is very helpful. Tomita created the fundamentals of the theory of nuclear magnetic resonance (NMR) absorption and relaxation for four-spin nuclear systems using the properties of the point symmetry of the molecular states in crystals by studying the intramolecular tunneling phenomenon in solid methane CH_4 [25]. Watton et al. [52, 53] have applied Tomita's approach to discuss NMR experiments in terms of the indistinguishability of ammonium protons in inorganic salts and the formation of symmetrical nuclear spin states: nuclear spin isomerism.

The symmetry properties of the classical rotation of molecules have been taken into consideration by applying the continuous symmetry groups of three-dimensional rotation by developing the rotational diffusion (RD) model [29, 40, 41]. O'Reilly and Tsang [54] used the point symmetry group approach by describing the temperature dependence of the proton magnetic spin-lattice relaxation times in powder ammonium sulfate $(NH_4)_2SO_4$. For the first time, the representation theory of the crystallographic point symmetry groups was used by solving the hindered molecular motion (HMM) problem in the framework of the specific angular jump model [27, 49]. The point group approach to solving the HMM problem was used in the framework of fixed angular jump (FAJ) model too [28, 50, 55]. Satisfactory fits of the experimental NMR relaxation rates in the single crystals NH_4Cl [27], $C_{10}H_{16}$, $C_{10}D_{16}$ [56], and $C_{12}D_{18}$ [57] were achieved by using the particular theories of HMM based on the point groups.

It is in the frameworks of general crystal physics laws that the symmetry of a molecule can differ from the symmetry of its site. Lowering the temperature then results in deforming the unit cell and hence distorting the local symmetry. This follows the violation of transformation properties of the molecular variables. The effect of local symmetry to spectroscopic parameters has been taken into consideration in vibrational spectroscopy [12] and electron paramagnetic resonance [58] by means of general methods of the group-theoretical analysis: the site-symmetrical approach [59], the factor-group technique [60], as well as the crystal-field theory [61, 62]. By using crystal-field theory, one could gain knowledge about the symmetry properties of both the molecular motion and the molecule site. However, this does not produce a solution to the general problem. For the same reason, methods of computer simulation are not useful here [63].

In order to achieve progress in HMM theory, we have suggested bringing together the progressive ideas underlying two alternative models—the RD model and the FAJ model—by modifying the initial distribution of the molecular vector orientations to a continuous and periodic one. This allows us to apply the Chapmen–Kolmogorov equation in finite differences by solving the stochastic problem of molecular jumps to the finite angles, as well as to use the representation theory of the point symmetry groups by accounting for the stationary states of the motion. Two point symmetries—the symmetry of the molecular motion and that of the molecule site—upgrade the microscopic level of the stochastic problem. As a result, we create an intermediate model of HMM called the extended angular jump model (EAJ model) [5, 35, 36, 64]. The main physical assumptions taken in the EAJ model are summarized below:

- The physical system consists of identical noncorrelated molecules, molecular fragments, or atomic groups for which HMM takes place by means of the angular rotational jumps of their intramolecular vectors.
- The HMM is a stationary random process described by a simple Markovian chain and obeys classical mechanics laws.
- The elementary jump angles form a crystallographic point symmetry group.
- The group elements falling into one class have equal probability.
- The site symmetry can be distinct from the motion symmetry.
- The probability density of the instantaneous orientation of any molecular vector has a continuous and periodic angular distribution in a long-term interval.

The following comments will clarify the statements presented above. In crystals, HMM occurs in relatively stable surroundings. Owing to the angular periodicity of interaction of the molecule with its neighbors, the orientation of a molecular vector shows a periodic distribution. At temperatures near absolute zero, the instantaneous direction of a molecular vector would coincide with the angular coordinate of the bottom of an orientational potential well. Thermal excitation enlarges the localization of the molecular vector within the limits of its potential well, and the angular distribution of the instantaneous direction of the molecular vector is to be described by a continuous and periodic function of probability density. This distribution is uniform in the RD model framework, and it is discrete and periodic in the FAJ model framework.

The average one-fold rotation angle of molecular vectors is equal to or is a multiple of the angles between the centers of any two potential wells permitted by symmetry. Thus, the hypothesis about a discontinuity of the reorientation angles adopted in the FAJ model persists also in the EAJ model. At the same time, a starting orientation of the molecular vector is arbitrary within the limits of the full solid angle 4π, as it is prescribed in the framework of the RD model. That is why the EAJ model is considered intermediate between the two antecedent models.

In isolated liquids and solids, the adiabatic approximation is usually justified when the majority of molecules occupy the ground electronic, vibrational, and rotational states. For a molecular vector displaying configuration equilibrium positions with high-energy barriers, the probability of overcoming a potential wall by means of a classical jump equals zero and the basic configuration state of the molecule becomes degenerate [39, 65–67]. The multiplicity of such degeneracy is equal to the number of equilibrium molecular configurations, which conforms to the order of the HMM point symmetry group.

If a molecular configuration is not equivalent among various molecular configurations with finite barriers to motion, this degeneracy is partially removed. If no molecular configuration is equivalent, the degeneracy prescribed by symmetry is completely removed. Application of the group-theoretical method allows us to establish the operator relation between various configurations and classify the hindered states of the molecular motion upon the irreducible representations of the HMM point symmetry group. In the framework of classical physics principles, the operators of rotation do not lead to the superposition of states with different symmetries of irreducible representations. Therefore, any stationary state of the molecular quantity symmetrized on the irreducible representation of the point symmetry group is an invariant of motion.

Thus, using the group-theoretical method, similarly to its quantum mechanical application, we can classify the stationary states of the classical HMM by the irreducible representations of a point symmetry group G.

It is useful to determine the statistical weight of a stationary state. For not very low temperatures ($T \geq 40$ K), there is no effect of nuclear spin isomerism on the classical motion. Hence, the symmetry of the nuclear spin states and the nuclear statistical weights can be neglected here [25, 52]. Consequently, within the limits of geometrical principles of invariance, the weight of a stationary hindered state $\langle \alpha, \nu |$ is equal to the dimension $\mu_\alpha^{(\nu)}$ of the irreducible representation $\Gamma_\alpha^{(\nu)}$, where α labels irreducible representations of rank ν.

The total weight of all states of the rank ν gives the weight of the irreducible representation $D^{(\nu)}$ of the continuous group of three-dimensional rotation O^{3+}. That is,

$$\sum_{\alpha} \mu_{\alpha}^{(\nu)} = 2\nu + 1 \tag{2.1}$$

Hence, the normalized weight of a degenerate state $\langle \alpha, \nu |$ denoted $q_{\alpha}^{(\nu)}$ is

$$q_{\alpha}^{(\nu)} = \mu_{\alpha}/(2\nu+1) \tag{2.2}$$

On the other hand, taking into consideration that the dimension μ_{α} is equal to the character $\chi_{\alpha E}$ of the identical class E in the representation $\Gamma_{\alpha}^{(\nu)}$, we can put it to use in an equivalent expression for $q_{\alpha}^{(\nu)}$

$$q_{\alpha}^{(\nu)} = \chi_{\alpha E}/(2\nu + 1) \tag{2.3}$$

By summing the weights $q_{\alpha}^{(\nu)}$ over the subindex α, we get their normalization condition

$$\sum_{\alpha} q_{\alpha}^{(\nu)} = 1 \tag{2.4}$$

The presented HMM symmetry properties are valid in the limits of the abstract group approach, and they agree well with the regular crystal lattices. In real crystals, symmetry is often not regular but is distorted. In order to take into account the effect of HMM on symmetry distortion, we suggest a phenomenology that a distortion of site symmetry results in deformation of the abstract group space followed by altering the weights $q_{\alpha}^{(\nu)}$. In addition, we assume that the symmetry elements prescribed by the abstract point group are also associate with the single steps of the real motion. It means that the group space does not vary and the normalization condition [Equation (2.4)] for the weights $q_{\alpha}^{(\nu)}$ remains. Therefore, we can give a physical meaning to a weight $q_{\alpha}^{(\nu)}$: the dynamical weight of the hindered state symmetrized on the irreducible representation $\Gamma_{\alpha}^{(\nu)}$ of the HMM symmetry group G. This specified property of the HMM symmetry, we interpret as the presence of the natural law of invariance in the molecular motion accounted for the symmetry of the interaction between the molecule and its surroundings [45].

Finally, we represent the main peculiarities of the EAJ model. A random physical rearrangement of individual atomic systems (molecules) takes place by means of proper and improper classical rotational displacements of the molecular vectors to finite angles between the instantaneous angular directions displaying a periodic distribution over the angles. The starting direction of a molecular vector is arbitrary in the limits of the full solid angle 4π. The maxima of the probability density of the instantaneous orientations fall into the angular equilibrium positions fixed on the center of potential wells. In a potential well, the molecule exhibits occasional attempts to overcome the energy barrier separating potential wells. Some attempts appear to be successful. An act of HMM is considered fulfilled if the molecular vector abandons the limits of the former potential well. A return of the molecular vector to the initial potential well incites an act to the HMM called an identical rotational jump. The angles of rotational jumps of a molecular vector form one of the 32 crystallographic point symmetry groups.

2.2 Solution of the stochastic problem

We shall start the solution of the stochastic problem by expanding the molecular quantity $F = F\,[\mathbf{b}(g)] = F(g)$, a function of the orientation g of the molecular vector **b**, on the full set of the basis functions $\Psi^{(\nu)}_{\alpha\beta}(g)$ of the HMM point symmetry group G. Fortunately, the functions $\Psi^{(\nu)}_{\alpha\beta}(g)$ are tabulated by Leushin [68] in the form of linear combinations of spherical harmonics up to the rank $\nu = 8$. Under a group rotation to the finite angle g_{jh}, the basis functions $\Psi^{(\nu)}_{\alpha\beta}(g)$ transform themselves by means of the operator of the finite rotational displacement $\mathbf{R}_\alpha(g_{jh})$ [65–67]

$$\Psi^{(\nu)}_{\alpha\beta}(g_h) = \mathbf{R}_\alpha(g_{jh})\Psi^{(\nu)}_{\alpha\beta}(g_j) = \sum_{s=1}^{\chi_{\alpha E}} \Gamma_{\alpha\beta s}(g_{jh})\Psi^{(\nu)}_{\alpha s}(g_j) \tag{2.5}$$

where $g_j = \theta_j, \varphi_j$ and $g_h = \theta_h, \varphi_h$ are two successive angular coordinates of the vector $\mathbf{b}(g) = \mathbf{b}(\theta, \varphi)$ determined in the polar reference frame (Figure 1.1). The quantity $\Gamma_{\alpha\beta s}(g_{jh})$ is an element of the transformation matrix $\Gamma_\alpha(g_{jh})$ of the irreducible representation $\Gamma^{(\nu)}_\alpha$, β and s label the lines and the columns of this matrix, $\chi_{\alpha E}$ is the character of the identical element E, and α labels the irreducible representations. The orthonormalization condition of the basis functions is

$$\oint \Psi^{(\nu)}_{\alpha\beta}(g)^* \Psi^{(\nu')}_{\alpha'\beta'}(g)\,\mathrm{d}g = \delta_{\alpha\alpha'}\,\delta_{\beta\beta'}\,\delta_{\nu\nu'} \tag{2.6}$$

where $\delta_{\alpha\alpha'}$, $\delta_{\beta\beta'}$, and $\delta_{\nu\nu'}$ are Kronecker deltas, the asterisk denotes complex conjugation, and $\mathrm{d}g = \sin\theta\,\mathrm{d}\theta\,\mathrm{d}\varphi$ is the element of solid angle.

According to the transformation rule given by Equation (2.5), the basis functions of various irreducible representations do not intermix under classical rotational displacement. This allows us to partition the solution of the stochastic problem into two steps. First, we solve the partial problem of rotational motion of a molecular vector **b** occupying the hindered state $\langle\alpha,\nu|$ symmetrized on the irreducible representation $\Gamma_\alpha^{(\nu)}$. Vector **b** is oriented by default in a fixed initial direction $\tilde{g}_0$. Then, we expand the result of this solution over all continuously distributed initial directions $\tilde{g}_0$ and discrete states $\langle\alpha,\nu|$.

Solving the partial stochastic problem for the state $\langle\alpha,\nu|$ is based on the Chapman–Kolmogorov probability equation in finite differences that links the probabilities of two successive discrete directions $\tilde{g}_j$ and $\tilde{g}_k$ of vector **b**

$$P_\alpha^{(\nu)}(\tilde{g}_0, N, \tilde{g}_k) = \sum_{j=1}^{\sigma} p(\tilde{g}_{jk}) P_\alpha^{(\nu)}(\tilde{g}_0, N-1, \tilde{g}_j) \tag{2.7}$$

Here, the quantity $P_\alpha^{(\nu)}(\tilde{g}_0, N-1, \tilde{g}_j)$ is the partial probability that vector **b** is oriented in the direction $\tilde{g}_j$ after exhibiting $N-1$ jumps from the initial direction $\tilde{g}_0$. Similarly, the quantity $P_\alpha^{(\nu)}(\tilde{g}_0, N, \tilde{g}_k)$ is the partial probability that this vector is oriented in the direction $\tilde{g}_k$ after exhibiting N jumps from the same initial direction $\tilde{g}_0$. The quantity $p(\tilde{g}_{jk})$ is the transition probability that vector **b** rotates from $\tilde{g}_j$ to $\tilde{g}_k$ by exhibiting a single rotational jump to the finite angle $\tilde{g}_{jk}$.

Let us determine the initial value of the partial probability density $P_\alpha^{(\nu)}(\tilde{g}_0, N, \tilde{g}_k)_{N=0} = P_\alpha^{(\nu)}(\tilde{g}_0, 0, \tilde{g}_k)$. Owing to the equivalence of the equilibrium configurations of the molecule, the transition probability denoted by $p(\tilde{g}_{jk})_i$ is the same for all single rotational displacements $p(\tilde{g}_{jk})$ of ith class. Denoting p_i, the probability of the molecular rearrangement by means of all elements of ith class, and σ_i, the order of this class, the relations between the jump probabilities are

$$\sum_{j=1}^{\sigma_i} p(\tilde{g}_{jk})_i = \sigma_i p(\tilde{g}_{jk})_i \equiv p_i \quad \text{or} \quad p(\tilde{g}_{jk})_i = \frac{p_i}{\sigma_i} \tag{2.8}$$

We replace Equation (2.7) with the help of Equation (2.8) by

$$P_\alpha^{(\nu)}(\tilde{g}_0, N, \tilde{g}_k) = \sum_i \sigma_i^{-1} p_i \sum_{j=1}^{\sigma_i} P_\alpha^{(\nu)}(\tilde{g}_0, N-1, \tilde{g}_j) \tag{2.9}$$

The solution of Equation (2.9) is performed by expanding the molecular quantities $P_\alpha^{(\nu)}(\tilde{g}_0, N-1, \tilde{g}_j)$ and $P_\alpha^{(\nu)}(\tilde{g}_0, N, \tilde{g}_k)$ in a series of the basis functions $\Psi_{\alpha\beta}^{(\nu)}(\tilde{g}_j)$ and $\Psi_{\alpha\beta}^{(\nu)}(\tilde{g}_k)$ of the motion point symmetry group G by

$$P_\alpha^{(\nu)}(\tilde{g}_0, N-1, \tilde{g}_j) = \sum_{\beta'} C_{\alpha\beta'}^{(\nu)}(\tilde{g}_0, N-1) \Psi_{\alpha\beta'}^{(\nu)}(\tilde{g}_j) \tag{2.10}$$

and

$$P_{\alpha}^{(\nu)}(\tilde{g}_0, N, \tilde{g}_k) = \sum_{\beta''} C_{\alpha\beta''}^{(\nu)}(\tilde{g}_0, N)\, \Psi_{\alpha\beta''}^{(\nu)}(\tilde{g}_k) \tag{2.11}$$

where β′ and β″ enumerate the lines of irreducible representations. By substituting the expansions given by Equations (2.10) and (2.11) in Equation (2.9), we obtain

$$\sum_{\beta''} C_{\alpha\beta''}^{(\nu)}(\tilde{g}_0, N)\Psi_{\alpha\beta''}^{(\nu)}(\tilde{g}_k) = \sum_{i} \sigma_i^{-1} p_i \sum_{\beta'} C_{\alpha\beta'}^{(\nu)}(\tilde{g}_0, N-1) \sum_{j=1}^{\sigma_i} \Psi_{\alpha\beta'}^{(\nu)}(\tilde{g}_j) \tag{2.12}$$

Functions $\Psi_{\alpha\beta'}^{(\nu)}(\tilde{g}_j)$ and $\Psi_{\alpha\beta''}^{(\nu)}(\tilde{g}_k)$ depend on the arguments $\tilde{g}_j$ and $\tilde{g}_k$, which determine the directions of vector **b** corresponding to two successive molecular configurations. Applying the transformation rule given by Equation (2.5) to the function $\Psi_{\alpha\beta'}^{(\nu)}(\tilde{g}_j)$, its dependence on the former argument $\tilde{g}_j$ transfers to the subsequent argument $\tilde{g}_k$:

$$\Psi_{\alpha\beta'}^{(\nu)}(\tilde{g}_j) = \mathbf{R}_{\alpha}\left(\tilde{g}_{jk}^{-1}\right)\Psi_{\alpha\beta'}^{(\nu)}(\tilde{g}_k) = \sum_{s=1}^{\chi_{\alpha E}} \Gamma_{\alpha\beta' s}\left(\tilde{g}_{jk}^{-1}\right)\Psi_{\alpha s}^{(\nu)}(\tilde{g}_k) \tag{2.13}$$

where $\mathbf{R}_{\alpha}(\tilde{g}_{jk}^{-1})$ is the operator of the inverse geometrical rotational displacement from orientation $\tilde{g}_k$ to orientation $\tilde{g}_j$, and $\Gamma_{\alpha\beta' s}(\tilde{g}_{jk}^{-1})$ is the β′s element of the matrix of inverse rotational displacement. In accordance with Schur's lemma [67], the sum of class matrix elements reduces to

$$\sum_{j=1}^{\sigma_i} \Gamma_{\alpha\beta' s}(\tilde{g}_{jk}^{-1}) = \sum_{j=1}^{\sigma_i} \Gamma_{\alpha\beta' s}(\tilde{g}_{jk}) = \sigma_i \chi_{\alpha i} \chi_{\alpha E}^{-1} \delta_{\beta' s} \tag{2.14}$$

where $\delta_{\beta' s}$ is Kronecker delta. With the help of Equations (2.13) and (2.14), the sum $\Sigma_{j=1}^{\sigma_i} \Psi_{\alpha\beta'}^{(\nu)}(\tilde{g}_j)$ transforms itself as

$$\begin{aligned}
\sum_{j=1}^{\sigma_i} \Psi_{\alpha\beta'}^{(\nu)}(\tilde{g}_j) &= \sum_{j=1}^{\sigma_i}\sum_{s=1}^{\chi_{\alpha E}} \Gamma_{\alpha\beta' s}\left(\tilde{g}_{jk}^{-1}\right)\Psi_{\alpha s}^{(\nu)}(\tilde{g}_k) = \sum_{s=1}^{\chi_{\alpha E}}\left[\sum_{j=1}^{\sigma_i} \Gamma_{\alpha\beta' s}\left(\tilde{g}_{jk}^{-1}\right)\right]\Psi_{\alpha s}^{(\nu)}(\tilde{g}_k) \\
&= \sum_{s=1}^{\chi_{\alpha E}}\left[\sum_{j=1}^{\sigma_i} \Gamma_{\alpha\beta' s}(\tilde{g}_{jk})\right]\Psi_{\alpha s}^{(\nu)}(\tilde{g}_k) = \sum_{s=1}^{\chi_{\alpha E}}\left[\sigma_i \chi_{\alpha i} \chi_{\alpha E}^{-1} \delta_{\beta' s}\right]\Psi_{\alpha s}^{(\nu)}(\tilde{g}_k) \\
&= \sigma_i \chi_{\alpha i} \chi_{\alpha E}^{-1} \sum_{s=1}^{\chi_{\alpha E}} \delta_{\beta' s} \Psi_{\alpha s}^{(\nu)}(\tilde{g}_k) = \frac{\sigma_i \chi_{\alpha i}}{\chi_{\alpha E}} \Psi_{\alpha\beta'}^{(\nu)}(\tilde{g}_k)
\end{aligned} \tag{2.15}$$

By using Equation (2.15), Equation (2.12) is replaced by

$$\begin{aligned}\sum_{\beta''} C^{(\nu)}_{\alpha\beta''}(\tilde{g}_0,N)\Psi^{(\nu)}_{\alpha\beta''}(\tilde{g}_k) &= \sum_i \sigma_i^{-1} p_i \sum_{\beta'} C^{(\nu)}_{\alpha\beta'}(\tilde{g}_0,N-1)\frac{\sigma_i \chi_{\alpha i}}{\chi_{\alpha E}}\Psi^{(\nu)}_{\alpha\beta'}(\tilde{g}_k) \\ &= \sum_i p_i \sum_{\beta'} \frac{\chi_{\alpha i}}{\chi_{\alpha E}} C^{(\nu)}_{\alpha\beta'}(\tilde{g}_0,N-1)\Psi^{(\nu)}_{\alpha\beta'}(\tilde{g}_k)\end{aligned} \tag{2.16}$$

We shall produce the orthogonalization procedure with Equation (2.16), multiplying both sides of Equation (2.16) by $\Psi^{(\nu)}_{\alpha\beta}(g_0)^*$ and integrating products upon full configuration space of the basis functions $\Psi^{(\nu)}_{\alpha\beta}(g_0)$:

$$\begin{aligned}&\oint \left[\sum_{\beta''} C^{(\nu)}_{\alpha\beta''}(\tilde{g}_0,N)\Psi^{(\nu)}_{\alpha\beta''}(\tilde{g}_k)\right]\Psi^{(\nu)}_{\alpha\beta'}(g_0)^* \mathrm{d}g_0 \\ &= \oint \sum_i p_i \sum_{\beta'} \frac{\chi_{\alpha i}}{\chi_{\alpha E}} C^{(\nu)}_{\alpha\beta'}(\tilde{g}_0,N-1)\Psi^{(\nu)}_{\alpha\beta'}(\tilde{g}_k)\Psi^{(\nu)}_{\alpha\beta'}(g_0)^* \mathrm{d}g_0\end{aligned} \tag{2.17}$$

Applying the orthonormalization condition given by Equation (2.6), we shall get a recurrence relation between the factors $C^{(\nu)}_{\alpha\beta}(\tilde{g}_0,N)$ and $C^{(\nu)}_{\alpha\beta}(\tilde{g}_0,N-1)$:

$$C^{(\nu)}_{\alpha\beta}(\tilde{g}_0,N) = \left(\chi_{\alpha E}^{-1}\sum_i \chi_{\alpha i} p_i\right) C^{(\nu)}_{\alpha\beta}(\tilde{g}_0,N-1) \tag{2.18}$$

By developing the factor $C^{(\nu)}_{\alpha\beta}(\tilde{g}_0,N-1)$ down to $N = 0$, we shall get a relation between the instantaneous and initial factors $C^{(\nu)}_{\alpha\beta}(\tilde{g}_0,N)$ and $C^{(\nu)}_{\alpha\beta}(\tilde{g}_0,0)$:

$$C^{(\nu)}_{\alpha\beta}(\tilde{g}_0,N) = \left(\chi_{\alpha E}^{-1}\sum_i \chi_{\alpha i} p_i\right)^N C^{(\nu)}_{\alpha\beta}(\tilde{g}_0,0) \tag{2.19}$$

This allows us to rewrite Equation (2.11) as

$$P^{(\nu)}_{\alpha}(\tilde{g}_0,N,\tilde{g}_k) = \sum_\beta \left(\chi_{\alpha E}^{-1}\sum_i \chi_{\alpha i} p_i\right)^N C^{(\nu)}_{\alpha\beta}(\tilde{g}_0,0)\,\Psi^{(\nu)}_{\alpha\beta}(\tilde{g}_k) \tag{2.20}$$

Equation (2.11) or (2.20) reduces for the initial instant, when $t = 0$ and $N = 0$, to

$$P^{(\nu)}_{\alpha}(\tilde{g}_0,0,\tilde{g}_k) = \sum_\beta C^{(\nu)}_{\alpha\beta}(\tilde{g}_0,0)\,\Psi^{(\nu)}_{\alpha\beta}(\tilde{g}_k) \tag{2.21}$$

In order to satisfy the initial condition of motion, the initial partial probability $P_{\alpha}^{(\nu)}(\tilde{g}_0, 0, \tilde{g}_k)$ is to be set as

$$P_{\alpha}^{(\nu)}(\tilde{g}_0, 0, \tilde{g}_k) = \frac{2\nu+1}{\mu_{\alpha}} \sum_{\beta} \left| \Psi_{\alpha\beta}^{(\nu)}(\tilde{g}_k) \right|^2 \delta_{0k} \tag{2.22}$$

or

$$P_{\alpha}^{(\nu)}(\tilde{g}_0, 0, \tilde{g}_k) = \frac{2\nu+1}{\mu_{\alpha}} \sum_{\beta} \Psi_{\alpha\beta}^{(\nu)}(\tilde{g}_0)^* \Psi_{\alpha\beta}^{(\nu)}(\tilde{g}_k) \delta_{0k} \tag{2.23}$$

From Equations (2.21) and (2.23), the initial value of the factor $C_{\alpha\beta}^{(\nu)}(\tilde{g}_0, 0)$ follows

$$C_{\alpha\beta}^{(\nu)}(\tilde{g}_0, 0) = \frac{2\nu+1}{\mu_{\alpha}} \Psi_{\alpha\beta}^{(\nu)}(\tilde{g}_0)^* \tag{2.24}$$

Substituting this value $C_{\alpha\beta}^{(\nu)}(\tilde{g}_0, 0)$ in Equation (2.20), we get the expression for the instantaneous partial probability $P_{\alpha}^{(\nu)}(\tilde{g}_0, N, \tilde{g}_k)$ as a function of the jump number N:

$$P_{\alpha}^{(\nu)}(\tilde{g}_0, N, \tilde{g}_k) = \frac{2\nu+1}{\mu_{\alpha}} \left(\chi_{\alpha E}^{-1} \sum_{i} \chi_{\alpha i} p_i \right)^N \sum_{\beta} \Psi_{\alpha\beta}^{(\nu)}(\tilde{g}_0)^* \, \Psi_{\alpha\beta}^{(\nu)}(\tilde{g}_k) \tag{2.25}$$

Using $w(N,t)$, the distribution function of the number of jumps by time, we can write a relation between the probability $P_{\alpha}^{(\nu)}(\tilde{g}_0, N, \tilde{g}_k)$ dependent on the number of jumps N and the probability $W_{\alpha}^{(\nu)}(\tilde{g}_0, t, \tilde{g}_k)$ dependent on time t. For a stationary Markovian process, this relation is [37]

$$W_{\alpha}^{(\nu)}(\tilde{g}_0, t, \tilde{g}_k) = \sum_{N=0}^{\infty} w(N,t) \, P_{\alpha}^{(\nu)}(\tilde{g}_0, N, \tilde{g}_k) \tag{2.26}$$

Function $w(N,t)$ is the probability that N successive molecular jumps occur during the time t. As usual, we shall accept the Poisson distribution for $w(N,t)$

$$w(N, t) = \frac{1}{N!} \left(\frac{t}{\tau} \right)^N \exp\left(-\frac{t}{\tau} \right) \tag{2.27}$$

where τ^{-1} is the distribution parameter and τ is the average time between two successive jumps.

After substituting Equations (2.25) and (2.27) into Equation (2.26) and convoluting the sum, we derive the expression for the time-dependent partial probability $W_{\alpha}^{(\nu)}(\tilde{g}_0,t,\tilde{g}_k)$, the solution to the HMM stochastic problem for the hindered state $\langle\alpha,\nu|$:

$$W_{\alpha}^{(\nu)}(\tilde{g}_0,t,\tilde{g}_k)=\sum_{\beta}\frac{2\nu+1}{\mu_{\alpha}}\Psi_{\alpha\beta}^{(\nu)}(\tilde{g}_0)^{*}\Psi_{\alpha\beta}^{(\nu)}(\tilde{g}_k)\exp\left(-\frac{t}{\tau_{\alpha}}\right) \tag{2.28}$$

where the dynamical parameter τ_{α}, the correlation time symmetrized on the irreducible representation Γ_{α}, is set by the following expression:

$$\tau_{\alpha}=\tau\left(1-\chi_{\alpha E}^{-1}\sum_{i}\chi_{\alpha i}p_i\right)^{-1} \tag{2.29}$$

Equation (2.28) determines a partial solution to the stochastic problem for a random molecular quantity, whose symmetry state is $\langle\alpha,\nu|$ and the initial value of the fixed angular argument is $\tilde{g}_0$. We shall find the final solution of the HMM problem by double-averaging Equation (2.28): first, over the various hindered states $\langle\alpha,\nu|$ with the help of the weight factors $q_{\alpha}^{(\nu)}$ and, second, over all initial directions $\tilde{g}_0$ with the help of Dirac delta function $\delta(\tilde{g}_0-g_0)$. Second averaging allows us to transform the discrete probability distribution into a continuous probability density distribution by taking into account the uniform distribution of $\tilde{g}_0$ within the full solid angle 4π. In other words, we get the binary probability density that the molecular vector, being directed in the limits of the unit solid angle along a direction g_0 at the initial instant, is oriented in the limits of the unit solid angle along an instantaneous direction $\tilde{g}_k$ at time t:

$$\begin{aligned}W(g_0,t,g_k)&=\frac{1}{4\pi}\oint\left[\sum_{\alpha,\nu}q_{\alpha}^{(\nu)}W_{\alpha}^{(\nu)}(\tilde{g}_0,t,\tilde{g}_k)\right]\delta(\tilde{g}_0-g_0)d\tilde{g}_0\\&=\frac{1}{4\pi}\sum_{\alpha,\nu}q_{\alpha}^{(\nu)}W_{\alpha}^{(\nu)}(g_0,t,\tilde{g}_k)\\&=\sum_{\alpha,\beta,\nu}q_{\alpha}^{(\nu)}\frac{2\nu+1}{4\pi\mu_{\alpha}}\Psi_{\alpha\beta}^{(\nu)}(g_0)^{*}\,\Psi_{\alpha\beta}^{(\nu)}(\tilde{g}_k)\,\exp\left(-\frac{t}{\tau_{\alpha}}\right).\end{aligned} \tag{2.30}$$

Because argument $\tilde{g}_k$ has a uniform distribution, the subscript k is not important in further consideration. Replacing μ_{α}, the dimension of the irreducible representation, by $\chi_{\alpha E}$, the character of the identity class, and omitting all signs marking the direction $\tilde{g}_k$, we get the final expression for the requested probability density in the general analytical form by

$$W(g_0,t,g)=\sum_{\alpha,\beta,\nu}\frac{2\nu+1}{4\pi\chi_{\alpha E}}q_{\alpha}^{(\nu)}\Psi_{\alpha\beta}^{(\nu)}(g_0)^{*}\Psi_{\alpha\beta}^{(\nu)}(g)\exp\left(-\frac{t}{\tau_{\alpha}}\right) \tag{2.31}$$

We demonstrate now the validity of this result to its initial conditions given by Equations (1.10) and (1.11).

Let us verify the validity of the first condition:

$$\begin{aligned}\oint W(g_0,0,g)dg &= \oint\sum_{\alpha,\beta,\nu}\frac{2\nu+1}{4\pi\chi_{\alpha E}}q_{\alpha}^{(\nu)}\Psi_{\alpha\beta}^{(\nu)}(g_0)^{*}\Psi_{\alpha\beta}^{(\nu)}(g)dg\\ &=\sum_{\alpha,\beta,\nu}\frac{2\nu+1}{4\pi\chi_{\alpha E}}q_{\alpha}^{(\nu)}\Psi_{\alpha\beta}^{(\nu)}(g_0)^{*}\oint\Psi_{\alpha\beta}^{(\nu)}(g)dg=\frac{1}{4\pi}\end{aligned} \tag{2.32}$$

Here we consider that the equalities given by Equations (2.33) and (2.34) are applicable when $\nu = 0$:

$$2\nu+1=1,\ q_{\alpha}^{(\nu=0)}=1,\ \chi_{\alpha E}=1,\ \Psi_{\alpha\beta}^{(\nu=0)}(g)=Y_0^{(0)}(\theta,\varphi)=\frac{1}{2\sqrt{\pi}} \tag{2.33}$$

and

$$\oint\Psi_{\alpha\beta}^{(\nu=0)}(g)dg=\int_{\theta=0}^{\pi}\int_{\varphi=0}^{2\pi}\frac{1}{2\sqrt{\pi}}\sin\theta\, d\theta\, d\varphi=2\sqrt{\pi} \tag{2.34}$$

whereas for $\nu \neq 0$, there is

$$\oint\Psi_{\alpha\beta}^{(\nu\neq0)}(g)dg=0 \tag{2.35}$$

Verifying the validity of the second condition:

$$\oint\!\!\oint W(g_0,0,g)\, dg\, dg_0=\oint\frac{1}{4\pi}dg_0=\frac{1}{4\pi}\int_0^{\pi}\int_0^{2\pi}\sin\theta\, d\theta\, d\varphi=1 \tag{2.36}$$

Thus, the verifications presented by Equations (2.32) and (2.36) are proof of the validity of solving the HMM problem for the EAJ model.

chapter three

Autocorrelation functions adapted to the extended angular jump model

3.1 General form of the autocorrelation functions

In order to calculate the normalized angular autocorrelation function (ACF) $K_m^{(\nu)}(t)$ given by Equation (1.13) with the help of Equation (2.31), the arguments $g_0 \equiv (\theta_0, \varphi_0)$ and $g \equiv (\theta, \varphi)$ of the spherical harmonics $Y_m^{(\nu)}(g_0)$ and $Y_m^{(\nu)}(g)^*$ as well as those for the basis functions $\Psi_{\alpha\beta}^{(\nu)}(g_0')^*$ and $\Psi_{\alpha\beta}^{(\nu)}(g')$ must be expressed in the same reference frame. Commonly, the arguments g_0 and g of functions $Y_m^{(\nu)}(g_0)$ and $Y_m^{(\nu)}(g)^*$ are given in the laboratory reference frame (LRF) and those for functions $\Psi_{\alpha\beta}^{(\nu)}(g_0')^*$ and $\Psi_{\alpha\beta}^{(\nu)}(g')$ in the crystallographic reference frame (CRF). Leushin [68] expresses the functions $\Psi_{\alpha\beta}^{(\nu)}(g')$ as the linear combination of the spherical harmonics $Y_n^{(\nu)}(g')$ by

$$\Psi_{\alpha\beta}^{(\nu)}(g') = \sum_{n=-\nu}^{\nu} \psi_{\alpha\beta n}^{(\nu)} \, Y_n^{(\nu)}(g') \tag{3.1}$$

where $\psi_{\alpha\beta n}^{(\nu)}$ are factors of decomposition tabulated up to $\nu = 8$.

With the help of Equation (1.7), the transformation rule of the functions $Y_n^{(\nu)}(g)$, we transform Equation (3.1) to the LRF by

$$\Psi_{\alpha\beta}^{(\nu)}(g') = \sum_{n,s=-\nu}^{\nu} \psi_{\alpha\beta n}^{(\nu)} \, D_{ns}^{(\nu)}(\Omega) \, Y_s^{(\nu)}(g) \tag{3.2}$$

where $D_{ns}^{(\nu)}(\Omega)$ is the rotation matrix element, Wigner function, and $\Omega = (\varphi, \vartheta, \xi)$ denotes Euler angles (Figure 1.2) of the single-crystal orientation in the LRF ($0 \le \varphi \le 2\pi$, $0 \le \vartheta \le \pi$, $0 \le \xi \le 2\pi$).

We shall get the general expression of ACF $K_m^{(\nu)}(\Omega,t)$ with the help of Equations (1.13), (2.31), and (3.2) by performing the following mathematical actions:

$$
\begin{aligned}
&K_m^{(\nu)}(\Omega,t)\\
&=\oiint Y_m^{(\nu)}(g_0)Y_m^{(\nu)}(g)^*\ W(g_0,t,g)\ dg_0 dg\\
&=\oiint Y_m^{(\nu)}(g_0)Y_m^{(\nu)}(g)^*\sum_{\alpha,\beta,\nu'}\frac{2\nu'+1}{4\pi\chi_{\alpha E}}q_\alpha^{(\nu')}\Psi_{\alpha\beta}^{(\nu')}(g_0')^*\,\Psi_{\alpha\beta}^{(\nu')}(g')\exp\left(-\frac{t}{\tau_\alpha}\right)dg_0dg\\
&=\oiint Y_m^{(\nu)}(g_0)Y_m^{(\nu)}(g)^*\sum_{\alpha,\beta,\nu'}\frac{2\nu'+1}{4\pi\chi_{\alpha E}}q_\alpha^{(\nu')}\sum_{n,s'=-\nu'}^{\nu'}\psi_{\alpha\beta n}^{(\nu')*}D_{ns'}^{(\nu')}(\Omega)^*\,Y_{s'}^{(\nu')}(g_0)^*\\
&\quad\cdot\sum_{n,s'=-\nu'}^{\nu'}\psi_{\alpha\beta n}^{(\nu')}D_{ns'}^{(\nu')}(\Omega)\,Y_{s'}^{(\nu')}(g)\exp\left(-\frac{t}{\tau_\alpha}\right)dg_0dg\\
&=\sum_{\alpha,\beta,\nu'}q_\alpha^{(\nu')}\frac{2\nu'+1}{4\pi\chi_{\alpha E}}\sum_{n,s'=-\nu'}^{\nu'}\psi_{\alpha\beta n}^{(\nu')}D_{ns'}^{(\nu')}(\Omega)\oint Y_m^{(\nu)}(g_0)Y_{s'}^{(\nu')}(g_0)^*dg_0\\
&\quad\cdot\sum_{n,s'=-\nu'}^{\nu'}\psi_{\alpha\beta n}^{(\nu')}D_{ns'}^{(\nu')}(\Omega)\oint Y_m^{(\nu)}(g)^*\,Y_{s'}^{(\nu')}(g)\ dg\exp\left(-\frac{t}{\tau_\alpha}\right)\\
&=\sum_{\alpha,\beta}q_\alpha^{(\nu)}\frac{2\nu+1}{4\pi\chi_{\alpha E}}\sum_{n=-\nu}^{\nu}\psi_{\alpha\beta n}^{(\nu)*}D_{nm}^{(\nu)}(\Omega)^*\sum_{n=-\nu}^{\nu}\psi_{\alpha\beta n}^{(\nu)}D_{nm}^{(\nu)}(\Omega)\exp\left(-\frac{t}{\tau_\alpha}\right)
\end{aligned}
\tag{3.3}
$$

We present the final expression of the required ACF in the form

$$
K_m^{(\nu)}(\Omega,t)=\frac{2\nu+1}{4\pi}\sum_{\alpha,\beta}\frac{q_\alpha^{(\nu)}}{\chi_{\alpha E}}\left|\sum_{n=-\nu}^{\nu}\psi_{\alpha\beta n}^{(\nu)}D_{nm}^{(\nu)}(\Omega)\right|^2\exp\left(-\frac{t}{\tau_\alpha}\right)
\tag{3.4}
$$

By averaging the monocrystalline ACF $K_m^{(\nu)}(\Omega,t)$ over the three-dimensional angle of single-crystal orientation $\Omega=\Omega\ (\varphi,\vartheta,\xi)$, we derive the powder ACF $K_m^{(\nu)}(t)$

$$K_m^{(\nu)}(t) = \left\langle K_m^{(\nu)}(\Omega,t)\right\rangle_\Omega = \frac{1}{8\pi^2}\oint K_m^{(\nu)}(\Omega,t)\,d\Omega$$

$$= \frac{1}{8\pi^2}\oint \frac{2\nu+1}{4\pi}\sum_{\alpha,\beta}\frac{q_\alpha^{(\nu)}}{\chi_{\alpha E}}\left|\sum_n \psi_{\alpha\beta n}^{(\nu)} D_{nm}^{(\nu)}(\Omega)\right|^2 \exp\left(-\frac{t}{\tau_\alpha}\right)d\Omega$$

$$= \frac{1}{4\pi}\sum_\alpha \frac{1}{\chi_{\alpha E}}\frac{2\nu+1}{8\pi^2}\oint\sum_{\beta,n}\left|\psi_{\alpha\beta n}^{(\nu)} D_{nm}^{(\nu)}(\Omega)\right|^2 d\Omega\cdot q_\alpha^{(\nu)}\exp\left(-\frac{t}{\tau_\alpha}\right)$$

$$= \frac{1}{4\pi}\sum_\alpha\left[\frac{1}{\chi_{\alpha E}}\sum_{\beta,n}\left|\psi_{\alpha\beta n}^{(\nu)}\right|^2\right]\cdot q_\alpha^{(\nu)}\exp\left(-\frac{t}{\tau_\alpha}\right) = \frac{1}{4\pi}\sum_\alpha q_\alpha^{(\nu)}\exp\left(-\frac{t}{\tau_\alpha}\right). \tag{3.5}$$

During calculations, the properties of orthogonal polynomials were used. The polycrystalline ACF expression is surprising: It has no dependence on subscript m

$$K_m^{(\nu)}(t) = K^{(\nu)}(t) = \frac{1}{4\pi}\sum_\alpha q_\alpha^{(\nu)}\exp\left(-\frac{t}{\tau_\alpha}\right). \tag{3.6}$$

The symmetrized correlation time τ_α is given by Equation (2.29). Sometimes, it is helpful to present τ_α by

$$\tau_\alpha = \tau(1-A_\alpha)^{-1} \tag{3.7}$$

where τ is the average time between two successive steps of the molecular motion, and A_α is the average one-fold probability transform of the molecular quantity upon the irreducible representation Γ_α. The expression for A_α is

$$A_\alpha = \chi_{\alpha E}^{-1}\sum_i p_i\,\chi_{\alpha i} \tag{3.8}$$

The hindered molecular motion (HMM) phenomenon is an activation process. Therefore, Arrhenius relation is valid in order to calculate the time interval τ:

$$\tau = \tau_0 \exp(E_a/RT) \tag{3.9}$$

Here E_a is the average height of the reorientation energy barrier and τ_0 is the average time interval between two successive attempts to overcome the barrier.

The numbering values of the theoretical parameters $\chi_{\alpha i}$, $\chi_{\alpha E}$, $\psi^{(v)}_{\alpha\beta n}$, $q^{(v)}_{\alpha}$, τ_α, E_a, and $D^{(v)}_{nm}(\Omega) \equiv D^{(v)}_{nm}(\phi, \vartheta, \xi)$ are important for practical application of Equations (3.4) through (3.9). One can find the values $\chi_{\alpha i}$, $\chi_{\alpha E}$, $\psi^{(v)}_{\alpha\beta n}$, and $D^{(v)}_{nm}(\phi, \vartheta, \xi)$, for instance, in the manuals [30, 38, 39, 68]. The symmetrized correlation times τ_α can be calculated by using Equations (2.29) and (3.7) through (3.9). The data for dynamical parameters τ_0, E_a, and p_i are unknown in advance. One has to determine them during computer simulation or experimentally, for instance, by studying the temperature and/or angular dependence of the symmetrized correlation times τ_α. As to the factors $q^{(v)}_{\alpha}$, the dynamical weights, they can be calculated for the regular crystalline structure by using Equation (2.3). However, in symmetry-distorted cases, there is no available explicit expression yet. Therefore, these factors become the fitted dynamical quantities in the HMM theory—the symmetrized dynamical weights of the hindered states. Determining the dynamical quantities τ_α and $q^{(v)}_{\alpha}$ is the major goal of the experimental spectroscopic studies with respect to the extended angular jump (EAJ) model.

3.2 *Autocorrelation functions of the first rank*

The first- and second-rank ACFs are mostly used in discussion of the experimental data presented by molecular spectroscopy techniques. In this section, we derive the explicit formulae of the first-rank ACFs adapted to the crystallographic point-symmetry groups of pure rotation. The elements of the crystallographic point-symmetry groups of pure rotation are pictured as required. The values of characters $\chi_{\alpha E}$ and $\chi_{\alpha i}$, the factors $\psi^{(v)}_{\alpha\beta n}$, and Wigner functions $D^{(v)}_{nm}(\Omega) \equiv D^{(v)}_{nm}(\phi, \vartheta, \xi)$ are given in Table 3.1 (Desks A–J) and Table 3.2 (Desks A–D) for $v = 1$ and 2. The numerical subscripts are used for the irreducible representations Γ_α in order to simplify further reference to them. Thus, $\alpha = 0$ labels identical representation. The non-identical representations are labeled by $\alpha = 1$, 2, and 3. The classification of the irreducible representations of the first rank and the respective explicit formulas of the average one-fold probability transforms A_α are given in Table 3.3.

Table 3.1. The Characters of the Irreducible Representations $\chi_{\alpha i}$ of the Crystallographic Point Symmetry Groups of Pure Rotation and the Factors $\psi^{(\nu)}_{\alpha\beta n}$ of Expansion of the Basis Functions $\Psi^{(\nu)}_{\alpha\beta}(g) = \Sigma^{\nu}_{n=-\nu}\psi^{(\nu)}_{\alpha\beta n}Y^{(\nu)}_{n}(g)$ on the Unitary Spherical Tensor Components $Y^{(\nu)}_{n}(g)$ of the Ranks $\nu = 1$ and 2

Desk A. Point symmetry group of the octahedron O

	$\chi_{\alpha i}$						$\psi^{(\nu)}_{\alpha\beta n}$				
$\Gamma_\alpha\backslash C_i$	1E	$6C_2$	$8C_3$	$6C_4$	$3C_4^2$	$\beta\backslash n$	−2	−1	0	1	2
						$\nu = 1$					
Γ_4, F_1	3	−1	0	1	−1	1		1	0	0	
						2		0	1	0	
						3		0	0	1	
						$\nu = 2$					
Γ_3, E	2	0	−1	0	2	1	0	0	1	0	0
						2	$2^{-1/2}$	0	0	0	$2^{-1/2}$
Γ_5, F_2	3	1	0	−1	−1	1	0	0	0	1	0
						2	$-i2^{-1/2}$	0	0	0	$i2^{-1/2}$
						3	0	1	0	0	0

Desk B. Point symmetry group of the tetrahedron T

	$\chi_{\alpha i}$					$\psi^{(\nu)}_{\alpha\beta n}$				
$\Gamma_\alpha\backslash C_i$	1E	$3C_2$	$4C_3^1$	$4C_3^2$	$\beta\backslash n$	−2	−1	0	1	2
					$\nu = 1$					
Γ_4', F	3	−1	0	0	1		1	0	0	
					2		0	1	0	
					3		0	0	1	
					$\nu = 2$					
Γ_{23}', E	2	2	−1	−1	1	0	0	1	0	0
					2	$2^{-1/2}$	0	0	0	$2^{-1/2}$
Γ_4', F	3	−1	0	0	1	0	1	0	0	0
					2	$-2^{-1/2}$	0	0	0	$2^{-1/2}$
					3	0	0	0	−1	0

(continued)

Table 3.1. The Characters of the Irreducible Representations $\chi_{\alpha i}$ of the Crystallographic Point Symmetry Groups of Pure Rotation and the Factors $\psi^{(\nu)}_{\alpha\beta n}$ of Expansion of the Basis Functions $\Psi^{(\nu)}_{\alpha\beta}(g) = \Sigma^{\nu}_{n=-\nu}\psi^{(\nu)}_{\alpha\beta n}Y^{(\nu)}_{n}(g)$ on the Unitary Spherical Tensor Components $Y^{(\nu)}_{n}(g)$ of the Ranks $\nu = 1$ and 2 (Continued)

Desk C. Point symmetry group D_4

	$\chi_{\alpha i}$						$\psi^{(\nu)}_{\alpha\beta n}$				
$\Gamma_\alpha \backslash C_i$	1E	$1C_2$	$2C_4$	$2U_2$	$2U_2'$	$\beta \backslash n$	−2	−1	0	1	2
						$\nu = 1$					
Γ_{t2}, A_2	1	1	−1	−1	1	1		0	1	0	
Γ_{t5}, **E**	2	−2	0	0	0	1		0	0	1	
						2		1	0	0	
						$\nu = 2$					
Γ_{t1}, $\mathbf{A_1}$	1	1	1	1	1	1	0	0	1	0	0
Γ_{t3}, $\mathbf{B_1}$	1	1	−1	1	−1	1	$2^{-1/2}$	0	0	0	$2^{-1/2}$
Γ_{t4}, $\mathbf{B_2}$	1	1	−1	−1	1	1	$-i2^{-1/2}$	0	0	0	$i2^{-1/2}$
Γ_{t5}, **E**	2	−2	0	0	0	1	0	0	0	−1	0
						2	0	1	0	0	0

Desk D. Point symmetry group C_4

	$\chi_{\alpha i}$					$\psi^{(\nu)}_{\alpha\beta n}$				
$\Gamma_\alpha \backslash C_i$	1E	$1C_2$	$1C_4$	$1C_4^3$	$\beta \backslash n$	−2	−1	0	1	2
						$\nu = 1$				
Γ_{t1}', **A**	1	1	1	−1	1		0	1	0	
Γ_{t34}', **E**	2	−2	0	0	1		0	0	1	
					2		1	0	0	
						$\nu = 2$				
Γ_{t1}', **A**	1	1	1	1	1	0	0	1	0	0
${}^1\Gamma_{t2}'$, ^{1}B	1	1	−1	−1	1	$2^{-1/2}$	0	0	0	$2^{-1/2}$
${}^2\Gamma_{t2}'$, ^{2}**B**	1	1	−1	−1	1	$-i2^{-1/2}$	0	0	0	$i2^{-1/2}$
Γ_{t34}', **E**	2	−2	0	0	1	0	0	0	−1	0
					2	0	1	0	0	0

Table 3.1. The Characters of the Irreducible Representations $\chi_{\alpha i}$ of the Crystallographic Point Symmetry Groups of Pure Rotation and the Factors $\psi^{(\nu)}_{\alpha\beta n}$ of Expansion of the Basis Functions $\Psi^{(\nu)}_{\alpha\beta}(g) = \Sigma^{\nu}_{n=-\nu}\psi^{(\nu)}_{\alpha\beta n}Y^{(\nu)}_{n}(g)$ on the Unitary Spherical Tensor Components $Y^{(\nu)}_{n}(g)$ of the Ranks $\nu = 1$ and 2 (Continued)

Desk E. Point symmetry group D_6

	$\chi_{\alpha i}$							$\psi^{(\nu)}_{\alpha\beta n}$				
$\Gamma_\alpha \backslash C_i$	1E	$1C_2$	$2C_3$	$2C_6$	$3U_2$	$3U_2'$	$\beta \backslash n$	−2	−1	0	1	2
							$\nu = 1$					
Γ_{h2}, **B**	1	1	1	1	−1	−1	1		0	1	0	
Γ_{h5}, $\mathbf{E_1}$	2	−2	−1	1	0	0	1		0	0	1	
							2		1	0	0	
							$\nu = 2$					
Γ_{h1}, $\mathbf{A_1}$	1	1	1	1	1	1	1	0	0	1	0	0
Γ_{h5}, $\mathbf{E_1}$	2	−2	−1	1	0	0	1	0	0	0	1	0
							2	0	−1	0	0	0
Γ_{h6}, $\mathbf{E_2}$	2	2	−1	−1	0	0	1	0	0	0	0	1
							2	1	0	0	0	0

Desk F. Point symmetry group C_6

	$\chi_{\alpha i}$					$\psi^{(\nu)}_{\alpha\beta n}$				
$\Gamma_\alpha \backslash C_i$	1E	$1C_2$	$2C_3$	$2C_6$	$\beta \backslash n$	−2	−1	0	1	2
					$\nu = 1$					
Γ_{h1}', **A**	1	1	1	1	1		0	i	0	
Γ_{h56}', $\mathbf{E_1}$	2	−2	−1	1	1		0	0	1	
					2		1	0	0	
					$\nu = 2$					
Γ_{h1}', **A**	1	1	1	1	1	0	0	1	0	0
Γ_{h34}', $\mathbf{E_1}$	2	2	−1	−1	1	0	0	0	0	1
					2	1	0	0	0	0
Γ_{h56}', $\mathbf{E_2}$	2	−2	−1	1	1	0	0	0	1	0
					2	0	−1	0	0	0

(*continued*)

Table 3.1. The Characters of the Irreducible Representations $\chi_{\alpha i}$ of the Crystallographic Point Symmetry Groups of Pure Rotation and the Factors $\psi^{(\nu)}_{\alpha\beta n}$ of Expansion of the Basis Functions $\Psi^{(\nu)}_{\alpha\beta}(g) = \Sigma^{\nu}_{n=-\nu}\psi^{(\nu)}_{\alpha\beta n}Y^{(\nu)}_{n}(g)$ on the Unitary Spherical Tensor Components $Y^{(\nu)}_{n}(g)$ of the Ranks $\nu = 1$ and 2 (Continued)

Desk G. Point symmetry group D_3

	$\chi_{\alpha i}$			$\psi^{(\nu)}_{\alpha\beta n}$					
$\Gamma_\alpha \backslash C_i$	1E	$2C_3$	$3U_2$	$\beta \backslash n$	−2	−1	0	1	2
				ν =1					
Γ_2^T, $\mathbf{A_2}$	1	1	−1	1		0	1	0	
Γ_3^T, **E**	2	−1	0	1		0	0	1	
				2		1	0	0	
				ν =2					
Γ_1^T, **A**	1	1	1	1	0	0	1	0	0
${}^1\Gamma_3^T$, ^{1}**E**	2	−1	0	1	0	0	0	1	0
				2	0	−1	0	0	0
${}^2\Gamma_3^T$, ^{2}**E**	2	−1	0	1	1	0	0	0	0
				2	0	0	0	0	1

Desk H. Point symmetry group C_3

	$\chi_{\alpha i}$			$\psi^{(\nu)}_{\alpha\beta n}$					
$\Gamma_\alpha \backslash C_i$	1E	1^1C_3	1^2C_3	$\beta \backslash n$	−2	−1	0	1	2
				ν =1					
$\Gamma_1'^T$, **A**	1	1	1	1		0	i	0	
$\Gamma_{23}'^T$, **E**	2	−1	−1	1		0	0	1	
				2		1	0	0	
				ν =2					
$\Gamma_1'^T$, **A**	1	1	1	1	0	0	1	0	0
${}^1\Gamma_{23}'^T$, ^{1}**E**	2	−1	−1	1	0	0	0	1	0
				2	0	−1	0	0	0
${}^2\Gamma_{23}'^T$, ^{2}**E**	2	−1	−1	1	1	0	0	0	0
				2	0	0	0	0	1

Table 3.1. The Characters of the Irreducible Representations $\chi_{\alpha i}$ of the Crystallographic Point Symmetry Groups of Pure Rotation and the Factors $\psi^{(\nu)}_{\alpha\beta n}$ of Expansion of the Basis Functions $\Psi^{(\nu)}_{\alpha\beta}(g) = \Sigma^{\nu}_{n=-\nu}\psi^{(\nu)}_{\alpha\beta n}Y^{(\nu)}_{n}(g)$ on the Unitary Spherical Tensor Components $Y^{(\nu)}_{n}(g)$ of the Ranks ν = 1 and 2 (Continued)

Desk I. Point symmetry group D_2

	$\chi_{\alpha i}$				$\psi^{(\nu)}_{\alpha\beta n}$					
$\Gamma_\alpha \backslash C_i$	1E	1^1C_2	1^2C_2	1^3C_2	β\n	−2	−1	0	1	2
						ν = 1				
Γ_{r2}, $\mathbf{B_1}$	1	−1	−1	1	1		$2^{-1/2}$	0	$2^{-1/2}$	
Γ_{r3}, $\mathbf{B_2}$	1	1	−1	−1	1		0	1	0	
Γ_{r4}, $\mathbf{B_3}$	1	−1	1	−1	1		$i2^{-1/2}$	0	$-i2^{-1/2}$	
						ν = 2				
$^1\Gamma_{r1}$, **A**	1	1	1	1	1	0	0	1	0	0
$^2\Gamma_{r1}$, **A**	1	1	1	1	1	$2^{-1/2}$	0	0	0	$2^{-1/2}$
Γ_{r2}, $\mathbf{B_1}$	1	−1	−1	1	1	0	$2^{-1/2}$	0	$-2^{-1/2}$	0
Γ_{r3}, $\mathbf{B_2}$	1	1	−1	−1	1	$-2^{-1/2}$	0	0	0	$2^{-1/2}$
Γ_{r4}, $\mathbf{B_3}$	1	−1	1	−1	1	0	$-i2^{-1/2}$	0	$-i2^{-1/2}$	0

Desk J. Point symmetry group C_2

	$\chi_{\alpha i}$		$\psi^{(\nu)}_{\alpha\beta n}$					
$\Gamma_\alpha \backslash C_i$	1E	$1C_2$	β\n	−2	−1	0	1	2
				ν = 1				
$^1\Gamma_{m1}$, **A**	1	1	1		0	i	0	
$^1\Gamma_{m2}$, ^{1}**B**	1	−1	1		$2^{-1/2}$	0	$2^{-1/2}$	
$^2\Gamma_{m2}$, ^{2}**B**	1	−1	1		$-i2^{-1/2}$	0	$i2^{-1/2}$	
				ν = 2				
$^1\Gamma_{m1}$, **A**	1	1	1	0	0	1	0	0
$^2\Gamma_{m1}$, **A**	1	1	1	$2^{-1/2}$	0	0	0	$2^{-1/2}$
$^3\Gamma_{m1}$, **A**	1	1	1	$-i2^{-1/2}$	0	0	0	$i2^{-1/2}$
$^1\Gamma_{m2}$, ^{1}**B**	1	−1	1	0	$2^{-1/2}$	0	$-2^{-1/2}$	0
$^2\Gamma_{m2}$, ^{2}**B**	1	−1	1	0	$i2^{-1/2}$	0	$i2^{-1/2}$	0

Table 3.2 Normalized Spherical Harmonics and Matrix Elements of Wigner of the First and Second Ranks

Desk A. Normalized first-rank spherical harmonics $Y_m^{(1)}(\theta,\varphi)$ ($m = -1, 0, 1$)

$$Y_{-1}^{(1)}(\theta,\varphi) = i\sqrt{3/8\pi}\,\sin\theta\, e^{-i\varphi}$$

$$Y_{0}^{(1)}(\theta,\varphi) = \sqrt{3/4\pi}\,\cos\theta$$

$$Y_{+1}^{(1)}(\theta,\varphi) = -i\sqrt{3/8\pi}\,\sin\theta\, e^{i\varphi}$$

Desk B. Matrix elements of Wigner (Wigner functions) of the first rank

$$D_{mn}^{(1)}(\Omega) \equiv D_{mn}^{(1)}(\phi,\vartheta,\xi) = e^{-im\xi} p_{mn}^{(1)}(\cos\vartheta) e^{-in\phi} \quad (m = -1, 0, 1;\ n = -1, 0, 1)$$

$p_{mn}^{(1)}(\cos\vartheta)$

$m \backslash n$	−1	0	1
−1	$(1+\cos\vartheta)/2$	$\sqrt{(1-\cos^2\vartheta)/2}$	$(1-\cos\vartheta)/2$
0	$-\sqrt{(1-\cos^2\vartheta)/2}$	$\cos\vartheta$	$\sqrt{(1-\cos^2\vartheta)/2}$
1	$(1-\cos\vartheta)/2$	$-\sqrt{(1-\cos^2\vartheta)/2}$	$(1+\cos\vartheta)/2$

Desk C. Normalized spherical harmonics of the second rank $Y_m^{(2)}(\theta,\varphi)$

$$Y_0^{(2)}(\theta,\varphi)=\sqrt{5/16\pi}\,(3\cos^2\theta-1)$$

$$Y_{\pm1}^{(2)}(\theta,\varphi)=\mp\sqrt{15/8\pi}\,\sin\theta\cos\theta e^{\pm i\varphi}$$

$$Y_{\pm2}^{(2)}(\theta,\varphi)=\sqrt{15/32\pi}\,\sin^2\theta\, e^{\pm 2i\varphi}$$

Desk D. Matrix elements of Wigner of the second rank $D_{mn}^{(2)}(\Omega)\equiv D_{mn}^{(2)}(\phi,\vartheta,\xi)$: $D_{mn}^{(2)}(\Omega)=e^{-im\xi}\,p_{mn}^{(2)}(\cos\vartheta)e^{-in\phi}$

$p_{mn}^{(2)}(\cos\vartheta)$

$m\backslash n$	−2	−1	0	1	2
−2	$(\cos\vartheta+1)^2/4$	$i\sin\vartheta(\cos\vartheta+1)/2$	$-(1-\cos^2\vartheta)\sqrt{(3/8)}$	$i\sin\vartheta(\cos\vartheta-1)/2$	$(\cos\vartheta-1)^2/4$
−1	$i\sin\vartheta(\cos\vartheta+1)/2$	$(2\cos^2\vartheta+\cos\vartheta-1)/2$	$i\sin\vartheta\cos\vartheta\sqrt{(3/2)}$	$(2\cos^2\vartheta-\cos\vartheta-1)/2$	$i\sin\vartheta(\cos\vartheta-1)/2$
0	$-(1-\cos^2\vartheta)\sqrt{(3/8)}$	$i\sin\theta\cos\vartheta\sqrt{(3/2)}$	$(3\cos^2\vartheta-1)/2$	$i\sin\vartheta\cos\vartheta\sqrt{(3/2)}$	$-(1-\cos^2\vartheta)\sqrt{(3/8)}$
1	$i\sin\theta(\cos\vartheta-1)/2$	$(2\cos^2\vartheta-\cos\vartheta-1)/2$	$i\sin\vartheta\cos\vartheta\sqrt{(3/2)}$	$(2\cos^2\vartheta+\cos\vartheta-1)/2$	$i\sin\vartheta(\cos\vartheta+1)/2$
2	$(\cos\vartheta-1)^2/4$	$i\sin\vartheta(\cos\vartheta-1)/2$	$-(1-\cos^2\vartheta)\sqrt{(3/8)}$	$i\sin\vartheta(\cos\vartheta+1)/2$	$(\cos\vartheta+1)^2/4$

Table 3.3 Classification of the Irreducible Representations $\Gamma_\alpha^{(1)}$ of the Crystallographic Point Symmetry Groups of Pure Rotation G of the First Rank and the Corresponding Explicit Expressions of the Average One-Fold Probability Transform A_α

Crystal system	G	α	$\Gamma_\alpha^{(1)}$	$A_\alpha = \chi_{\alpha E}^{-1} \sum_i p_i \chi_i$
Cubic	O	1	Γ_4, F_1	$p(E) - p(C_2)/3 + p(C_4)/3 - p(C_4^2)/3$
	T	1	Γ_4', F	$p(E) - p(C_2)/3$
Tetragonal	D_4	1	Γ_{t5}, E	$p(E) - p(C_2)$
		2	Γ_{t2}, A_2	$p(E) + p(C_2) + p(C_4) - p(U_2) - p(U_2')$
	C_4	0	Γ_{t1}', A	1
		1	Γ_{t34}', E	$p(E) - p(C_2)$
Hexagonal	D_6	1	Γ_{h5}, E_1	$p(E) - p(C_2) - p(C_3)/2 + p(C_6)/2$
		2	Γ_{h2}, B	$p(E) + p(C_2) + p(C_3) + p(C_6) - p(U_2) - p(U_2')$
	C_6	0	Γ_{h1}', A	1
		1	Γ_{h56}', E_1	$p(E) - p(C_2) - p(C_3)/2 + p(C_6)/2$
Trigonal	D_3	1	Γ_3^T, E	$p(E) - p(C_3)/2$
		2	Γ_2^T, A_2	$p(E) + p(C_3) - p(U_2)$
	C_3	0	$\Gamma_1'^T$, A	1
		1	$\Gamma_{23}'^T$, E	$p(E) - p(C_3)/2$
Orthorhombic	D_2	1	Γ_{r2}, B_1	$p(E) - p(^1C_2) - p(^2C_2) + p(^3C_2)$
		2	Γ_{r4}, B_3	$p(E) - p(^1C_2) + p(^2C_2) - p(^3C_2)$
		3	Γ_{r3}, B_2	$p(E) + p(^1C_2) - p(^2C_2) - p(^3C_2)$
Monoclinic	C_2	0	Γ_{m1}, A	1
		1	$^1\Gamma_{m2}$, ^1B	$p(E) - p(C_2)$
		2	$^2\Gamma_{m2}$, ^2B	$p(E) - p(C_2)$

3.2.1 *Point symmetry groups of the cubic systems O and T*

The symmetry elements of the pure rotation point groups of octahedron O and tetrahedron T are shown respectively in Figure 3.1(a) and Figure 3.1(b). The following notations are taken to five classes of the group O: E corresponds to the identity class, C_2 corresponds to the class of six (two-fold) rotations of angle π about the diagonal axes of cube faces passing through the opposite edge centers, C_3 corresponds to the class of eight (three-fold) rotations of angles $\pm 2\pi/3$ about the axes making body diagonals of the cube. Each of two classes C_4 and C_4^2 contains three (four-fold and two-fold) rotations, respectively, of angles $\pi/2$ and π about the axes passing through the opposite face centers.

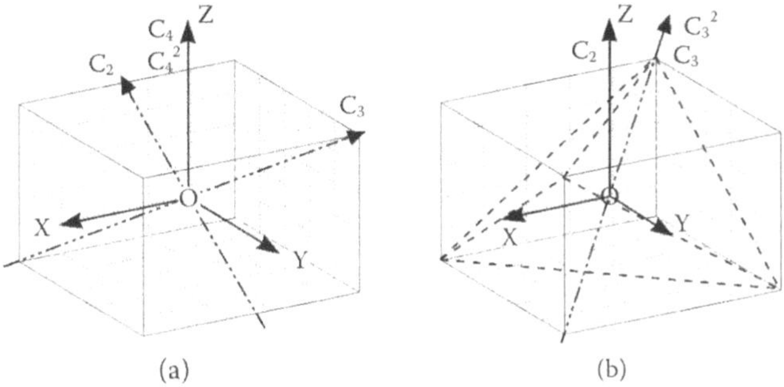

Figure 3.1 Elements of the pure rotation crystallographic point symmetry groups of cubic system: (a) the group of octahedron O; (b) the group of tetrahedron T. Only one element of every nonidentical symmetry class is shown.

In group T, E corresponds to the identity class; C_2 to the class of three (two-fold) rotations of angle π about the axes passing through opposite edge centers of the tetrahedron; and C_3^1 and C_3^2 to two classes with four respective (three-fold) rotations of angles $2\pi/3$ and $4\pi/3$ about the axes passing through the corners of the tetrahedron and their opposite face centers. Frequently, a unified class C_3 of eight rotations of angles $\pm 2\pi/3$ replaces two classes C_3^1 and C_3^2.

Regarding the data presented in Table 3.1 (Desks A and B) and Table 3.3, we can see that the unique first-rank three-dimensional irreducible representation corresponds to both groups O and T denoted respectively by Γ_4 (or F_1) and Γ_4' (or F). This irreducible representation and the physical quantities symmetrized on this representation are labeled by the numerical index $\alpha = 1$. The unitary spherical tensor components of the first rank $Y_n^{(1)}(\theta,\varphi)$, with $n = -1$, 0, and 1, present the basis functions for both groups. Therefore, all factors $\psi_{\alpha\beta n}^{(1)}$ of Equation (3.1) are equal to 1. The characters of the identity class E are equal to $\chi_{1E} = \chi(\Gamma_4)_E = \chi(\Gamma_4')_E = 3$. Denoting the value of the single correlation time τ_α by τ_1, and accounting for the factors $\psi_{1,1,-1}^{(1)} = \psi_{1,2,0}^{(1)} = \psi_{1,3,1}^{(1)} = 1$ and the character of identity class $\chi_{1E} = 3$, the expression of ACF given by Equation (3.4) takes the form

$$K_m^{(1)}(\Omega,t) = \frac{1}{4\pi} q_1^{(1)} \left(\sum_{n=-1}^{1} \left| D_{nm}^{(1)}(\Omega) \right|^2 \right) \exp\left(-\frac{t}{\tau_1} \right). \tag{3.10}$$

We can simplify Equation (3.10). Because of the uniqueness of the irreducible representation $\Gamma_\alpha = \Gamma_1$, its dynamical weight is equal to $q_\alpha^{(1)} \equiv q_1^{(1)} = 1$. In addition, the sum of squares of Wigner functions is normalized by $\Sigma_{n=-1}^{1} | D_{nm}^{(1)}(\Omega) |^2 = 1$. Accounting for these numbering data, the explicit expression of $K_m^{(1)}(\Omega, t)$ reduces at last to $K^{(1)}(t)$,

$$K^{(1)}(t) = \frac{1}{4\pi} \exp\left(-\frac{t}{\tau_1} \right) \tag{3.11}$$

which has no dependence on Ω and m.

By using the class characters presented in Table 3.1 (Desks A and B), we can express the correlation times $\tau_1(\mathrm{O})$ and $\tau_1(\mathrm{T})$ adapted to the groups O and T with respect to Equation (2.29) by

$$\tau_1(\mathrm{O}) \equiv \tau(\Gamma_4) = \tau\,[1 - p(\mathrm{E}) + p(\mathrm{C}_2)/3 - p(\mathrm{C}_4)/3 + p(\mathrm{C}_4^2)/3]^{-1} \tag{3.12}$$

and

$$\tau_1(\mathrm{T}) \equiv \tau(\Gamma_4') = \tau\,[1 - p(\mathrm{E}) + p(\mathrm{C}_2)/3]^{-1} \tag{3.13}$$

According to Equation (3.11), the first-rank ACF adapted to the groups of cubic systems is exponential and isotropic. Hence the same formula is also available for the first-rank angular ACF in powder samples.

3.2.2 *Point groups of the axial symmetry C_n ($n = 3, 4, 6$)*

The symmetry elements of the groups C_3, C_4, and C_6 are shown in Figure 3.2(a), Figure 3.3(a), and Figure 3.4(a), respectively. According to the data of Table 3.1 (Desks D, F, and H) and Table 3.3, every group includes two irreducible representations: an identical one-dimensional A ($\alpha = 0$) and nonidentical two-dimensional E ($\alpha = 1$). It can also be seen that the basis functions are the same for all groups C_n ($n = 3, 4, 6$). Therefore, the respective pre-exponential factors of these ACFs will be equally expressed.

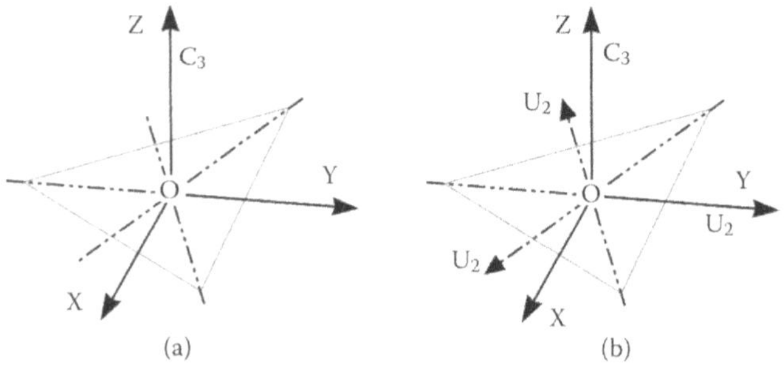

Figure 3.2 Elements of the pure rotation crystallographic point symmetry groups of the trigonal system: (a) group C_3; (b) group D_3.

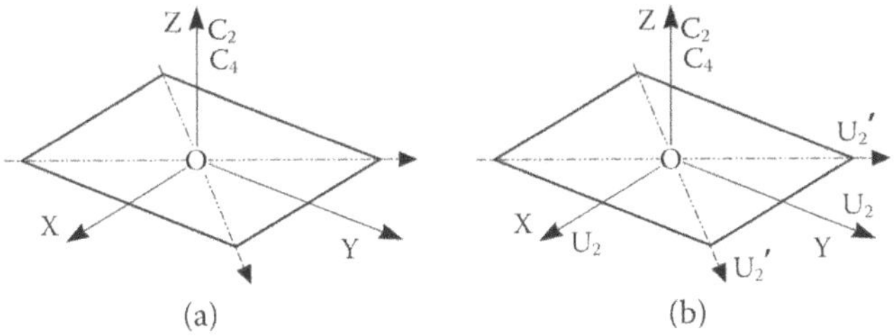

Figure 3.3 Elements of the pure rotation crystallographic point symmetry groups of the tetragonal system: (a) group C_4; (b) group D_4.

Now we present systematic calculation of the ACFs adapted to groups C_n ($n = 3, 4, 6$). The term of ACF symmetrized on the identical representation A does not vary with time because its average one-fold probability transform equals to $A_0 = 1$ (see Table 3.3), and, with respect to Equation (3.6), the correlation time is infinitely long ($\tau_{\alpha=0} = \infty$). The dynamical weight of identical representation is denoted by $q_A^{(1)} = q_0$. The decreasing part of ACF is symmetrized on the nonidentical representation E, and the respective dynamical weight and correlation time is denoted by $q_E^{(1)} = q_1$ and $\tau_E = \tau_1$.

Equation (3.4), the general expression of ACF, adapted to the groups C_n, takes the form:

$$K_m^{(1)}(\Omega,t) = \frac{3}{4\pi}\left\{ q_0 \left|\psi_{010}^{(1)} D_{0m}^{(1)}(\Omega)\right|^2 + \frac{q_1}{2}\left[\left|\psi_{111}^{(1)} D_{1m}^{(1)}(\Omega)\right|^2 + \left|\psi_{12-1}^{(1)} D_{-1m}^{(1)}(\Omega)\right|^2\right] \exp\left(-\frac{t}{\tau_1}\right)\right\} \tag{3.14}$$

The numbering values of factors $\psi_{\alpha\beta n}^{(1)}$ and explicit expressions of Wigner functions $D_{mn}^{(1)}(\Omega)$ are given in Table 3.1 (Desk D) and Table 3.2

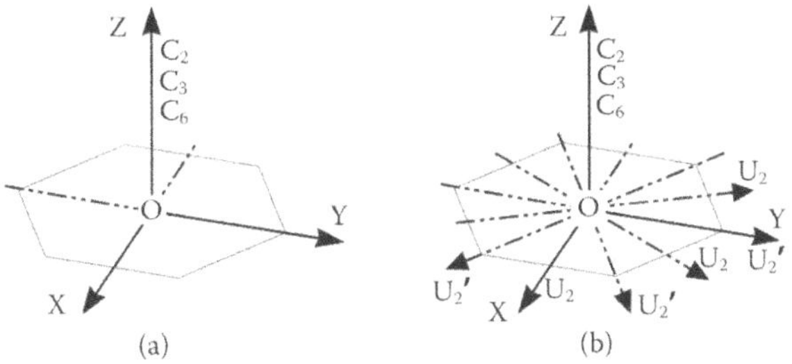

Figure 3.4 Elements of the pure rotation crystallographic point symmetry groups of the hexagonal system: (a) group C_6; (b) group D_6.

(Desk B). They are presented in an appropriate form below:

$$\psi^{(1)}_{0,1,0} = \psi^{(1)}_{1,1,1} = \psi^{(1)}_{1,2,-1} = 1,$$

$$\left|D^{(1)}_{0,0}(\Omega)\right| = \cos^2\vartheta,$$

$$\left|D^{(1)}_{1,0}(\Omega)\right| = \left|D^{(1)}_{-1,0}(\Omega)\right| = \left|D^{(1)}_{0,1}(\Omega)\right| = \left|D^{(1)}_{0,-1}(\Omega)\right| = \sqrt{(1-\cos^2\vartheta)/2},$$

$$\left|D^{(1)}_{1,-1}(\Omega)\right| = \left|D^{(1)}_{-1,1}(\Omega)\right| = (1-\cos\vartheta)/2,$$

$$\left|D^{(1)}_{1,1}(\Omega)\right| = \left|D^{(1)}_{-1,-1}(\Omega)\right| = (1+\cos\vartheta)/2. \tag{3.15}$$

By substituting these data in Equation (3.14), we get the desired ACF expressions

$$\begin{aligned} K^{(1)}_0(\Omega,t) &= K^{(1)}_0(\vartheta,t) \\ &= \frac{3}{4\pi}\left\{q_0\left|\psi^{(1)}_{010}D^{(1)}_{00}(\Omega)\right|^2 + \frac{1}{2}q_1\left[\left|\psi^{(1)}_{111}D^{(1)}_{10}(\Omega)\right|^2 + \left|\psi^{(1)}_{12-1}D^{(1)}_{-10}(\Omega)\right|^2\right]\exp\left(-\frac{t}{\tau_1}\right)\right\} \\ &= \frac{3}{4\pi}\left\{q_0\cos^2\vartheta + \frac{1}{2}q_1\left[\frac{(1-\cos^2\vartheta)}{2} + \frac{(1-\cos^2\vartheta)}{2}\right]\exp\left(-\frac{t}{\tau_1}\right)\right\} \\ &= \frac{3}{8\pi}\left[2q_0\cos^2\vartheta + q_1\sin^2\vartheta\exp\left(-\frac{t}{\tau_1}\right)\right] \end{aligned} \tag{3.16}$$

and

$$K^{(1)}_1(\Omega,t) = K^{(1)}_1(\vartheta,t) = K^{(1)}_{-1}(\Omega,t) = K^{(1)}_{-1}(\vartheta,t) = K^{(1)}_{\pm1}(\Omega,t) = K^{(1)}_{\pm1}(\vartheta,t)$$

$$= \frac{3}{4\pi}\left\{q_0\left|\psi^{(1)}_{110}D^{(1)}_{01}(\Omega)\right|^2 + \frac{1}{2}q_1\left[\left|\psi^{(1)}_{111}D^{(1)}_{11}(\Omega)\right|^2 + \left|\psi^{(1)}_{12-1}D^{(1)}_{-11}(\Omega)\right|^2\right]\exp\left(-\frac{t}{\tau_1}\right)\right\}$$

$$= \frac{3}{4\pi}\left\{ q_0 \frac{(1-\cos^2\vartheta)}{2} + \frac{1}{2} q_1 \left[\frac{(1+\cos\vartheta)^2}{4} + \frac{(1-\cos\vartheta)^2}{4} \right] \exp\left(-\frac{t}{\tau_1} \right) \right\}$$

$$= \frac{3}{4\pi}\left[\frac{1}{2} q_0 \sin^2\vartheta + \frac{1}{4} q_1 (1+\cos^2\vartheta) \exp\left(-\frac{t}{\tau_1} \right) \right]$$

$$= \frac{3}{16\pi}\left[2q_0 \sin^2\vartheta + q_1 (1+\cos^2\vartheta) \exp\left(-\frac{t}{\tau_1} \right) \right]. \tag{3.17}$$

It is useful to rewrite side by side the resulting explicit expressions $K_0^{(1)}(\vartheta,t)$ and $K_{\pm 1}^{(1)}(\vartheta,t)$ adapted to the groups C_n ($n = 3, 4, 6$)

$$K_0^{(1)}(\vartheta,t) = \frac{3}{8\pi}\left[2q_0 \cos^2\vartheta + q_1 \sin^2\vartheta \exp\left(-\frac{t}{\tau_1} \right) \right] \tag{3.18}$$

and

$$K_{\pm 1}^{(1)}(\vartheta,t) = \frac{3}{16\pi}\left[2q_0 \sin^2\vartheta + q_1 (1+\cos^2\vartheta) \exp\left(-\frac{t}{\tau_1} \right) \right] \tag{3.19}$$

The powder expression of the ACF adapted to the groups C_n ($n = 3, 4, 6$) equals in accordance with Equation (3.5) to

$$K^{(1)}(t) = \frac{1}{4\pi}\left[q_0 + q_1 \exp\left(-\frac{t}{\tau_1} \right) \right] \tag{3.20}$$

The characters of classes $\chi_{\alpha i}$ are presented in Table 3.2 (Desks D, F, and H) and the correlation time τ_1 calculated by using Equation (2.29) relative to various groups C_n ($n = 3, 4, 6$) are given by the following formulas:

Group C_3:

$$\tau_1(C_3) = \tau\,[1 - p(E) + p(C_3)/2]^{-1} \tag{3.21}$$

with $p(C_3) = p(C_3^1) + p(C_3^2)$.

Group C_4:

$$\tau_1(C_4) = \tau\,[1 - p(E) + p(C_2)]^{-1} \tag{3.22}$$

Group C_6:

$$\tau_1(C_6) = \tau\,[1 - p(E) + p(C_2)/2 - p(C_3)/2 + p(C_6)/2]^{-1} \tag{3.23}$$

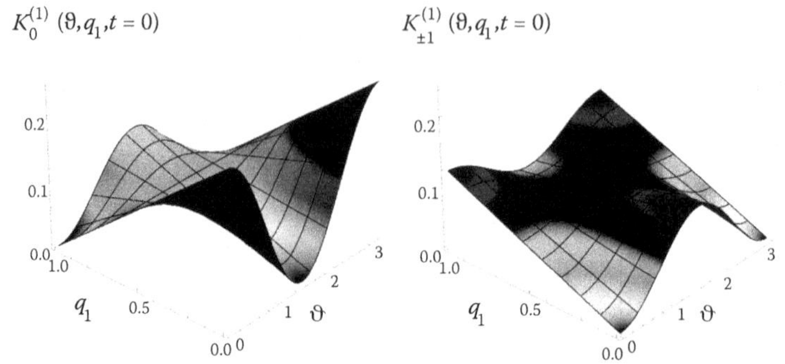

Figure 3.5 Graphs of the ACF peak values $K_0^{(1)}(\vartheta, q_1, t = 0)$, left, and $K_{\pm 1}^{(1)}(\vartheta, q_1, t = 0)$, right, adapted to the point symmetry groups C_n and D_n ($n = 3, 4, 6$). They are drawn as functions of the polar angle ϑ of the CRF orientation in the LRF and the dynamical weight of the two-dimensional irreducible representation q_1.

In Equations (3.18) through (3.20), the weighted parameters q_0 and q_1 satisfy the normalization condition (2.4). Therefore, they can take values from 0 to 1. Accounting for the data $\nu = 1$, $\chi_{0E} = 1$, and $\chi_{1E} = 2$, the static weights of the hindered states calculated by using Equation (3.3) take the values $q_0 = 1/3$ and $q_1 = 2/3$.

Thus, an ACF adapted to the groups C_n ($n = 3, 4, 6$) consists of the sum of two terms: one of them has no dependence on time and other one is exponentially decreasing with time. In single crystals, the ACF shows the orientation dependence on the polar angle ϑ. At the same time, there is no dependence on the azimuthal angle φ for these ACFs. Surface plot graphs of the peak values of the ACFs $K_0^{(1)}(\vartheta, q_1, 0)$ and $K_1^{(1)}(\vartheta, q_1, 0) = K_{-1}^{(1)}(\vartheta, q_1, 0)$ are shown as functions of the polar angle ϑ and the dynamical weight q_1 in Figure 3.5. We can see that the peak value of ACFs varies by changing the dynamical weights.

3.2.3 *Point symmetry groups D_n ($n = 3, 4, 6$)*

The symmetry elements of the groups D_3, D_4, and D_6 are shown in Figure 3.2(b), Figure 3.3(b), and Figure 3.4(b), respectively. The representation $D^{(1)}$ of the rank $\nu = 1$, being irreducible in the continuous rotation group O^{3+}, reduces to two nonidentical, irreducible representations labeled by $\alpha = 1$ and 2: two-dimensional E and one-dimensional A_2 on the groups D_3 and D_4 or two-dimensional E and one-dimensional B on the group D_6 (Table 3.3). The dynamical weights $q_1^{(1)} = q_1$ and $q_2^{(1)} = q_2$ satisfy the normalization condition $q_1 + q_2 = 1$.

Regarding Table 3.1 (Desks C, E, and G), the characters $\chi_{\alpha E}$ of the equal dimension representations and the factors $\psi_{\alpha\beta n}^{(1)}$ are the same for all groups D_n ($n = 3, 4, 6$). Performing the similar processing as for the groups C_n, we get the explicit expressions of the first-rank ACF adapted to the groups D_n ($n = 3, 4, 6$) by

$$K_0^{(1)}(\vartheta,t) = \frac{3}{8\pi}\left[q_1 \sin^2\vartheta \exp\left(-\frac{t}{\tau_1}\right) + 2q_2 \cos^2\vartheta \exp\left(-\frac{t}{\tau_2}\right)\right] \tag{3.24}$$

and

$$K_{\pm1}^{(1)}(\vartheta,t) = \frac{3}{16\pi}\left[q_1(1+\cos^2\vartheta)\exp\left(-\frac{t}{\tau_1}\right) + 2q_2 \sin^2\vartheta \exp\left(-\frac{t}{\tau_2}\right)\right] \tag{3.25}$$

In powder samples, they reduce to

$$K^{(1)}(t) = \frac{1}{4\pi}\left[q_1 \exp\left(-\frac{t}{\tau_1}\right) + q_2 \exp\left(-\frac{t}{\tau_2}\right)\right] \tag{3.26}$$

The respective correlation times τ_1 and τ_2 equal to

$$\tau_1(D_3) = \tau\,[1 - p(E) + p(C_3)/2]^{-1} \tag{3.27}$$

$$\tau_2(D_3) = \tau\,[1 - p(E) - p(C_3) + p(U_2)]^{-1} \tag{3.28}$$

$$\tau_1(D_4) = \tau[1 - p(E) + p(C_2)]^{-1} \tag{3.29}$$

$$\tau_2(D_4) = \tau\,[1 - p(E) - p(C_2) - p(C_4) + p(U_2) + p(U_2')]^{-1} \tag{3.30}$$

$$\tau_1(D_6) = \tau\,[1 - p(E) + p(C_2) + p(C_3)/2 - p(C_6)/2]^{-1} \tag{3.31}$$

$$\tau_2(D_6) = \tau\,[1 - p(E) - p(C_2) - p(C_3) - p(C_6) + p(U_2) + p(U_2')]^{-1} \tag{3.32}$$

The ACFs adapted to the groups D_n ($n = 3, 4, 6$) consist generally of two exponentially decreasing terms symmetrized on two-dimensional and one-dimensional representations. If one of the dynamical weights q_1 or q_2 equals 0, the ACF is exponential. The graphs of the peak values of single-crystalline ACFs $K_0^{(1)}(\vartheta, q_1, t=0)$ and $K_{\pm1}^{(1)}(\vartheta, q_1, t=0)$ adapted to the groups D_n ($n = 3, 4, 6$) are the same as for groups C_n ($n = 3, 4, 6$), displayed in Figure 3.5. We can see that the shape of anisotropy varies by changing the weight q_1. If weights q_1 and q_2 are equal to their static values—that is, respectively, to 2/3 and 1/3—the anisotropy of the ACF peak values vanishes. The straight lines parallel to the axis ϑ map this effect in Figure 3.5.

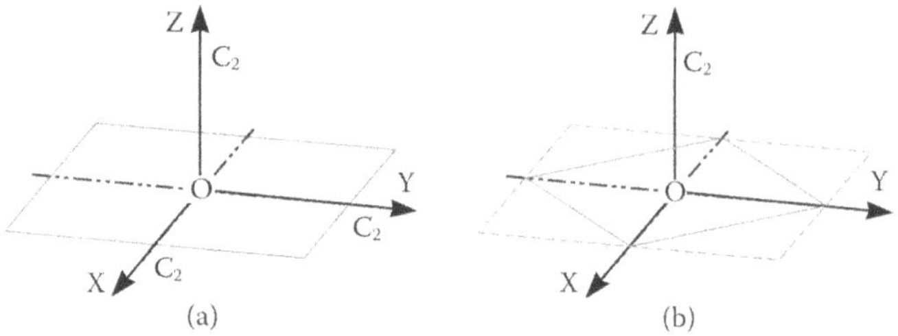

Figure 3.6 Elements of the pure rotation crystallographic point symmetry groups: (a) group D_2; (b) group C_2.

3.2.4 *Point symmetry group D_2*

The symmetry elements of the point group D_2 are shown in Figure 3.6(a). There are three one-dimensional nonequivalent, nonidentical irreducible representations ($\alpha = 1, 2, 3$) in this group. The dynamical weights $q_1^{(1)} = q_1$, $q_2^{(1)} = q_2$, and $q_3^{(1)} = q_3$ satisfy the normalization condition $q_1 + q_2 + q_3 = 1$.

The ACF expressions derived for this group by using Equations (3.4) and (3.5) with the help of the data $\psi_{\alpha\beta n}^{(1)}$, $\chi_{\alpha E}$, and $D_{nm}^{(1)}(\phi,\vartheta,\xi)$ [taken from Table 3.1 (Desk I) and Table 3.2] are given by Equations (3.33) through (3.35). They consist of three exponential terms. The ACFs depend on both angles ϕ and ϑ of the crystal orientation. The single-crystalline ACFs are

$$K_0^{(1)}(\vartheta,\phi,t) = \frac{3}{4\pi}\left[q_1 \cos^2\phi \sin^2\vartheta \exp\left(-\frac{t}{\tau_1}\right)\right.$$
$$\left. q_2 \sin^2\phi \sin^2\vartheta \exp\left(-\frac{t}{\tau_2}\right) + q_3 \cos^2\vartheta \exp\left(-\frac{t}{\tau_3}\right)\right] \tag{3.33}$$

and

$$K_{\pm1}^{(1)}(\vartheta,\phi,t) = \frac{3}{8\pi}\left[q_1(\cos^2\phi + \sin^2\phi\cos^2\vartheta)\exp\left(-\frac{t}{\tau_1}\right)\right.$$
$$\left. + q_2(\sin^2\phi + \cos^2\phi\cos^2\vartheta)\exp\left(-\frac{t}{\tau_2}\right) + q_3 \sin^2\vartheta \exp\left(-\frac{t}{\tau_3}\right)\right] \tag{3.34}$$

The expression of polycrystalline ACF $K_m^{(1)}(t) = K^{(1)}(t)$ is equal to

$$K^{(1)}(t) = \frac{1}{4\pi}\left[q_1 \exp\left(-\frac{t}{\tau_1}\right) + q_2 \exp\left(-\frac{t}{\tau_2}\right) + q_3 \exp\left(-\frac{t}{\tau_3}\right)\right] \tag{3.35}$$

The time constants τ_1, τ_2, and τ_3 are expressed as

$$\tau_1(D_2) = \tau\,[1 - p(E) + p(^1C_2) + p(^2C_2) - p(^3C_2)]^{-1} \tag{3.36}$$

$$\tau_2(D_2) = \tau\,[1 - p(E) + p(^1C_2) - p(^2C_2) + p(^3C_2)]^{-1} \tag{3.37}$$

and

$$\tau_3(D_2) = \tau\,[1 - p(E) - p(^1C_2) + p(^2C_2) + p(^3C_2)]^{-1} \tag{3.38}$$

Because of the multiplicity of variable parameters (the dynamical weights q_1, q_2, q_3, and the angles ϕ and ϑ), it is difficult to show the full graphs of the peak values for the single-crystalline ACFs predicted by Equations (3.33) and (3.34). It should be noted that in the case of equality of dynamical weights to their static values $q_1 = q_2 = q_3 = 1/3$, there is no anisotropy of the ACF peak values. Generally, the considered ACFs are anisotropic. For instance, the graphs of the peak value of ACFs $K_0^{(1)}(\phi,\vartheta,t=0)$ and $K_{\pm1}^{(1)}(\phi,\vartheta,t=0)$ are shown as functions of the azimuth ϕ and polar ϑ angles of the CRF orientation in the LRF for group D_2 in Figure 3.7, where the values of dynamical weights are fixed by $q_1 = 0.2$, $q_2 = 0.5$, and $q_3 = 0.3$.

The experimental data of the ACF peak value measured for some fixed angles of crystal orientation allow us to determine the numbering values of dynamical weights. Thus, by measuring $K_0^{(1)}(\phi,\vartheta,t=0)$ for $\vartheta = 0$, we can unambiguously determine the weight q_3. If $\phi = 0$ and $\vartheta = \pi/2$, the value $K_0^{(1)}(\phi,\vartheta,t=0)$ depends only on q_1. At last, if $\phi = \pi/2$ and $\vartheta = \pi/2$,

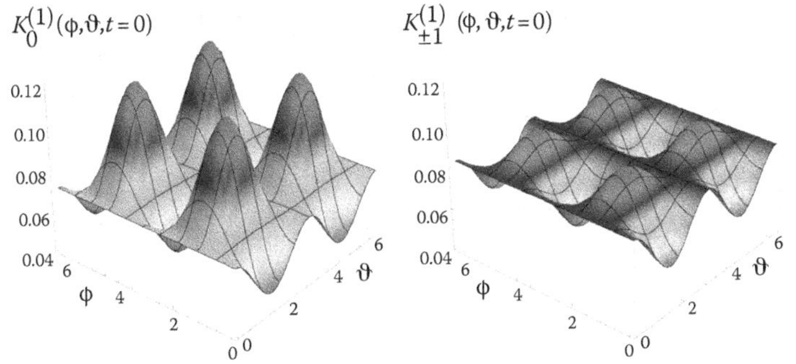

Figure 3.7 Graphs of the ACF peak values $K_0^{(1)}(\phi,\vartheta,t=0)$, left, and $K_{\pm1}^{(1)}(\phi,\vartheta,t=0)$, right, adapted to the point symmetry group D_2. They are drawn as functions of the azimuthal angle ϕ and the polar angle ϑ of the CRF orientation in the LRF. The fitted parameters of the dynamical weights are taken equal to $q_1 = 0.2$, $q_2 = 0.5$, and $q_3 = 0.3$.

it depends only on q_2. Formulas relating the dynamical weights to the selected peak values of $K_0^{(1)}(\phi,\vartheta,t=0)$ are

$$q_1 = (4\pi/3)\ K_0^{(1)}\ (\phi = 0,\ \vartheta = \pi/2,\ t = 0) \tag{3.39}$$

$$q_2 = (4\pi/3)\ K_0^{(1)}\ (\phi = \pi/2,\ \vartheta = \pi/2,\ t = 0) \tag{3.40}$$

and

$$q_3 = (4\pi/3)\ K_0^{(1)}\ (\vartheta = 0,\ t = 0) \tag{3.41}$$

3.2.5 *Point symmetry group* C_2.

The symmetry elements of group C_2 are shown in Figure 3.6(b). The data $\psi_{\alpha\beta n}^{(1)}$, $\chi_{\alpha E}$, and $D_{nm}^{(1)}(\phi,\vartheta,\xi)$ are presented in Table 3.1 (Desk J) and Table 3.2. The representation $D^{(1)}$ of the group O^{3+} reduces into an identical A ($\alpha = 0$) and two nonidentical, one-dimensional equivalent representations of group C_2 (Table 3.3): ^{1}B ($\alpha = 1$) and ^{2}B ($\alpha = 2$).

The ACF expressions derived for this group by using Equations (3.4) and (3.6) are given by Equations (3.42) through (3.44). The single-crystal-line ACFs depend on angles ϕ and ϑ of the crystal orientation

$$K_0^{(1)}(\vartheta,\phi,t) = \frac{3}{4\pi}\left[q_0\cos^2\vartheta + \left(q_1\sin^2\phi + q_2\cos^2\phi\right)\sin^2\vartheta \exp\left(-\frac{t}{\tau_1}\right)\right] \tag{3.42}$$

and

$$K_{\pm1}^{(1)}(\vartheta,\phi,t) = \frac{3}{8\pi}\left\{q_0\sin^2\vartheta + \left[q_1(\cos^2\phi + \sin^2\phi\cos^2\vartheta) + q_2(\sin^2\phi + \cos^2\phi\cos^2\vartheta)\right]\exp\left(-\frac{t}{\tau_1}\right)\right\} \tag{3.43}$$

where $q_1 \equiv q_1^{(1)}$, $q_2 \equiv q_2^{(1)}$, and $q_0 \equiv q_0^{(1)}$.

In powder samples, the ACFs reduce to a single expression

$$K^{(1)}(t) = \frac{1}{4\pi}\left[q_0 + (q_1 + q_2)\exp\left(-\frac{t}{\tau_1}\right)\right] \tag{3.44}$$

The equal geometrical subspaces correspond to two equivalent irreducible representations ^{1}B and ^{2}B. However, there is no evidence regarding the physical equivalence of the respective HMM states. Therefore, each pre-exponential factor of the ACFs displays two terms symmetrized on

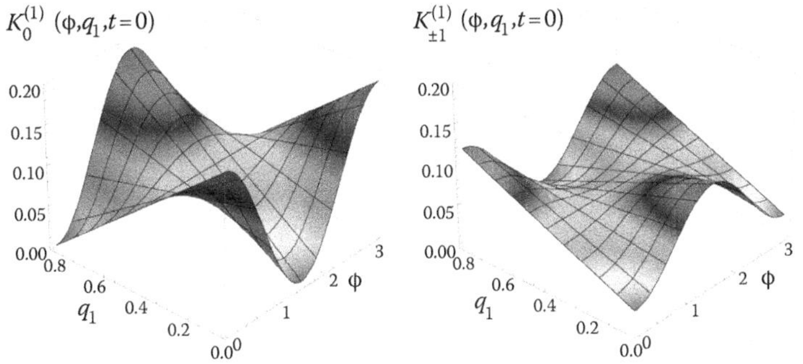

Figure 3.8 Graph of the ACF peak values $K_0^{(1)}(\phi, q_1, t = 0)$, left, and $K_{\pm1}^{(1)}(\phi, q_1, t = 0)$, right, adapted to the HMM point symmetry group C_2, drawn as functions of ϕ, the azimuthal angle of the CRF orientation in the LRF, and the dynamical weight q_1. The fitted parameters are equal to $q_0 = 0.2$ and $\vartheta = \pi/2$.

various equivalent irreducible representations. The single symmetrized correlation time derived by using Equation (2.29) equals

$$\tau_1 = \tau\,[1 - p(E) + p(C_2)]^{-1} \tag{3.45}$$

The graphs of the ACF peak values given by Equations (3.42) and (3.43) are shown in Figure 3.8 through Figure 3.10. For determinacy, the dynamical weight of the hindered state symmetrized on the identical irreducible representation is taken equal to $q_0 = 0.2$. In this case, the dynamical weights q_1 and q_2 satisfy the condition $q_1 + q_2 = 0.8$; that is, the weight q_1 as well as the weight q_2 can take values from 0 up to 0.8 by giving 0.8 as the sum.

The graph $K_m^{(1)}(\phi, q_1, t = 0)$, where $m = 0$ and ±1, is drawn as a function of the azimuthal angle ϕ and the dynamical weight q_1 for the fixed polar

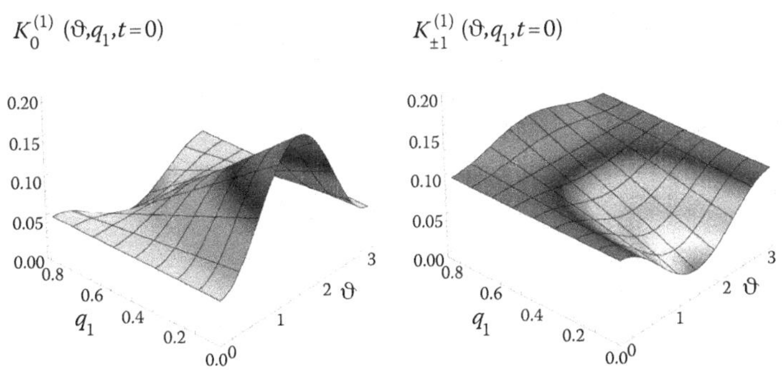

Figure 3.9 Graphs of the ACF peak values $K_0^{(1)}(\vartheta, q_1, t = 0)$, left, and $K_{\pm1}^{(1)}(\vartheta, q_1, t = 0)$, right, adapted to the HMM point symmetry group C_2, drawn as functions of ϑ, the polar angle of the CRF orientation in the LRF, and the dynamical weight q_1. The fitted parameters are equal to $q_0 = 0.2$ and $\phi = 0$.

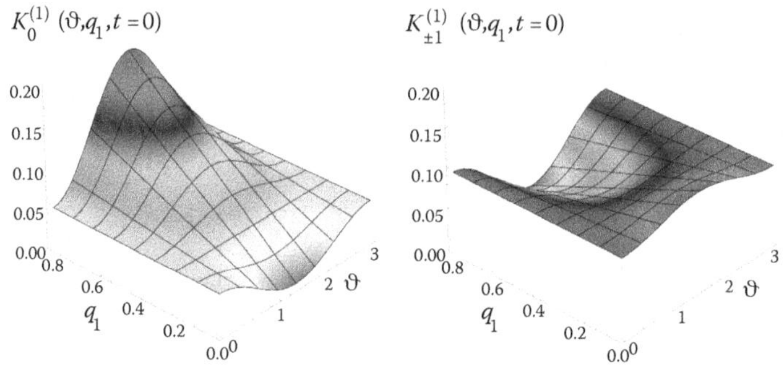

Figure 3.10 Graphs of the ACF peak values $K_0^{(1)}(\vartheta, q_1, t=0)$, left, and $K_{\pm1}^{(1)}(\vartheta, q_1, t=0)$, right, adapted to the point symmetry group C_2, drawn as functions of ϑ, the polar angle of the CRF orientation in the LRF, and the dynamical weight q_1. The fitted parameters are equal to $q_0 = 0.2$ and $\phi = \pi/2$.

angle $\vartheta = \pi/2$ of the CRF orientation with respect to the LRF in Figure 3.8. In cases when the irreducible representations ^{1}B and ^{2}B are not physically equivalent, a two-dimensional anisotropy of the ACFs takes place. If $q_1 = q_2$, the equivalent representations are equivalent physically and the ACF dependence of the azimuth angle ϕ vanishes. The straight line, parallel to axis ϕ, maps this theoretical outcome in Figure 3.8.

The graphs drawn for two fixed values of the azimuth $\phi = 0$ and $\phi = \pi/2$ are shown respectively in Figure 3.9 and Figure 3.10. In these cases, Equation (3.42) reduces for $\phi = 0$ to

$$K_0^{(1)}(\vartheta, t=0) = \frac{3}{4\pi}\left(q_0 \cos^2\vartheta + q_2 \sin^2\vartheta\right) \tag{3.46}$$

and for $\phi = \pi/2$ to

$$K_0^{(1)}(\vartheta, t=0) = \frac{3}{4\pi}\left(q_0 \cos^2\vartheta + q_1 \sin^2\vartheta\right) \tag{3.47}$$

The dependences (3.46) and (3.47) allow us to determine the respective numbering values of weights q_2 and q_1.

3.3 *Autocorrelation functions of the arbitrary rank*

Similar to the explicit presentation of the first-rank ACFs, we give the analytical expression of arbitrary rank symmetry-adapted ACFs by

$$K_m^{(\nu)}(\vartheta, \phi, t) = \frac{2\nu+1}{4\pi}\sum_{\alpha}\sum_{l=0}^{\nu} q_\alpha^{(\nu)} a_{\alpha l m}^{(\nu)}(\phi)\cos^{2l}\vartheta \exp\left(-\frac{t}{\tau_\alpha}\right) \tag{3.48}$$

where the factors $a^{(\nu)}_{\alpha lm}(\phi)$ are functions of sine or cosine of the azimuthal angle ϕ; ϑ is the polar angle of the CRF main axis orientation in the LRF, and summation by α includes summation of all irreducible representations, inclusive of identical representations.

In conformity to Equation (3.48), the working expressions for the first (ν = 1, l = 0 and 1) and second (ν = 2, l = 0, 1, 2) rank ACFs respectively are

$$K^{(1)}_m(\vartheta,\phi,t)=\frac{3}{4\pi}\sum_{\alpha}\sum_{l=0,1}q^{(1)}_{\alpha}a^{(1)}_{\alpha lm}(\phi)\cos^{2l}\vartheta\,\exp\left(-\frac{t}{\tau_\alpha}\right) \tag{3.49}$$

and

$$K^{(2)}_m(\vartheta,\phi,t)=\frac{5}{4\pi}\sum_{\alpha}\sum_{l=0,1,2}q^{(2)}_{\alpha}a^{(2)}_{\alpha lm}(\phi)\cos^{2l}\vartheta\,\exp\left(-\frac{t}{\tau_\alpha}\right) \tag{3.50}$$

The explicit expressions of the factors $a^{(1)}_{\alpha lm}(\phi)$ and $a^{(2)}_{\alpha lm}(\phi)$ are presented in Table 3.4 through Table 3.7. With respect to Equation (3.7), the symmetrized correlation times are the functions of the average one-fold probability transform A_α. The explicit expressions A_α are presented in Table 3.3 and Table 3.8.

The surface-plot graphs of the peak values of the first-rank ACFs are shown in Figure 3.5, and Figure 3.7 through Figure 3.10. The examples of the graphs of the peak values of the second-rank ACFs adapted to the groups of cubic, hexagonal-trigonal, and tetragonal crystallographic systems are shown in Figure 3.11 through Figure 3.15. In Figure 3.11 and

Table 3.4 Explicit Form of the Factors $a^{(1)}_{\alpha lm}(\phi)$ of the First-Rank ACFs Adapted to the Crystallographic Point Symmetry Groups of Pure Rotation

Point	α	1		2		3	
group	$m\backslash l$	0	1	0	1	0	1
O, T	0	1/3					
	1	1/3					
C_3, C_4, C_6	0	1/2	−1/2				
	1	1/4	1/4				
D_3, D_4, D_6	0	1/2	−1/2	0	1		
	1	1/4	1/4	1/2	−1/2		
C_2	0	$(\cos^2\phi)/2$	$-(\cos^2\phi)/2$	$(\sin^2\phi)/2$	$-\sin^2\phi)/2$		
	1	$(\cos^2\phi)/2$	$(\sin^2\phi)/2$	$(\sin^2\phi)/2$	$(\cos^2\phi)/2$		
D_2	0	$(\cos^2\phi)/2$	$-(\cos^2\phi)/2$	$(\sin^2\phi)/2$	$-(\sin^2\phi)/2$	0	1
	1	$(\cos^2\phi)/2$	$(\sin^2\phi)/2$	$(\sin^2\phi)/2$	$(\cos^2\phi)/2$	1/2	−1/2

Table 3.5 Explicit Form of the Factors $a^{(2)}_{\alpha l m}(\phi)$ of the Second-Rank ACFs Adapted to the Pure Rotational Point Symmetry Groups of Cubic, Hexagonal, and Trigonal Crystallographic Systems

Crystal system	ϕ	α / $m \backslash l$	1: 0	1: 1	1: 2	2: 0	2: 1	2: 2
Cubic	ϕ	0	$(1/8)(1+3\cos^2 2\phi)$	$-(3/4)(1+\cos^2 2\phi)$	$(3/8)(3+\cos^2 2\phi)$	$(1/4)(1-\cos^2 2\phi)$	$(1/2)(1+\cos^2 2\phi)$	$-(1/4)(3+\cos^2 2\phi)$
		1	$(1/4)(1-\cos^2 2\phi)$	$(1/2)(1+\cos^2 2\phi)$	$-(1/4)(3+\cos^2 2\phi)$	$(1/6)(1+\cos^2 2\phi)$	$-(1/3)(1+\cos^2 2\phi)$	$(1/6)(3+\cos^2 2\phi)$
		2	$(1/16)(3+\cos^2 2\phi)$	$-(1/8)(1+\cos^2 2\phi)$	$(1/16)(3+\cos^2 2\phi)$	$(1/24)(5-\cos^2 2\phi)$	$(1/12)(1+\cos^2 2\phi)$	$-(1/24)(3+\cos^2 2\phi)$
	0	0	1/2	−3/2	3/2	0	1	−1
		1	0	1	−1	1/3	−2/3	2/3
		2	1/4	−1/4	1/4	1/6	1/6	−1/6
	$\pi/4$	0	1/8	−3/4	9/8	1/4	1/2	−3/4
		1	1/4	1/2	−3/4	1/6	−1/3	1/2
		2	3/16	−1/8	3/16	5/24	1/12	−1/8
Hexagonal and Trigonal	Arbitrary	0	3/8	−3/4	3/8	0	3/2	−3/2
		1	1/4	0	−1/4	1/4	−3/4	1
		2	1/16	3/8	1/16	1/4	0	−1/4

Table 3.6 Explicit Form of the Factors $a^{(2)}_{\alpha l m}(\phi)$ of the Second-Rank ACFs Adapted to the Pure Rotational Point Symmetry Groups of Tetragonal Crystallographic System

ϕ	α	1			2			3		
	$m\backslash l$	0	1	2	0	1	2	0	1	2
ϕ	0	$(3/4)\cos^2 2\phi$	$-(3/2)\cos^2 2\phi$	$(3/4)\cos^2 2\phi$	$(3/4)\sin^2 2\phi$	$-(3/2)\sin^2 2\phi$	$(3/4)\sin^2 2\phi$	0	3/2	–3/2
	1	$(1/2)\sin^2 2\phi$	$(1/2)\cos 4\phi$	$-(1/2)\cos^2 2\phi$	$(1/2)\cos^2 2\phi$	$-(1/2)\cos 4\phi$	$-(1/2)\sin^2 2\phi$	1/4	–3/4	1
	2	$(1/8)\cos^2 2\phi$	$(1/4)(1+\sin^2 2\phi)$	$(1/8)\cos^2 2\phi$	$(1/8)\sin^2 2\phi$	$(1/4)(1+\cos^2 2\phi)$	$(1/8)\sin^2 2\phi$	1/4	0	–1/4
0	0	3/4	–3/2	3/4	0	0	0	0	3/2	–3/2
	1	0	1/2	–1/2	1/2	–1/2	0	1/4	–3/4	1
	2	1/8	1/4	1/8	0	1/2	0	1/4	0	–1/4
$\pi/4$	0	0	0	0	3/4	–3/2	3/4	0	3/2	–3/2
	1	1/2	–1/2	0	0	1/2	–1/2	1/4	–3/4	1
	2	0	1/2	0	1/8	1/4	1/8	1/4	0	–1/4

Table 3.7 Explicit Form of the Factors $a^{(2)}_{\alpha l m}(\phi)$ of the Second-Rank ACFs Adapted to the Pure Rotational Point Symmetry Groups of Orthorhombic and Monoclinic Crystallographic Systems

ϕ	α	1			2			3		
	$m\backslash l$	0	1	2	0	1	2	0	1	2
Arbitrary	0	0	$3\sin^2\phi$	$-3\sin^2\phi$	0	$3\cos^2\phi$	$-3\cos^2\phi$	$(3/4)\sin^2 2\phi$	$-(3/2)\sin^2 2\phi$	$(3/4)\sin^2 2\phi$
	1	$(1/2)\sin^2\phi$	$(1/2)(1-5\sin^2\phi)$	$2\sin^2\phi$	$(1/2)\cos^2\phi$	$(1/2)(1-5\cos^2\phi)$	$2\cos^2\phi$	$(1/2)\cos^2 2\phi$	$-(1/2)\cos 4\phi$	$-(1/2)\sin^2 2\phi$
	2	$(1/2)\cos^2\phi$	$-(1/2)\cos 2\phi$	$-(1/2)\sin^2\phi$	$(1/2)\sin^2\phi$	$(1/2)\cos 2\phi$	$-(1/2)\cos^2\phi$	$(1/8)\cos^2 2\phi$	$(1/4)(1+\cos^2 2\phi)$	$(1/8)\cos^2 2\phi$
0	0	0	0	0	0	3	−3	0	0	0
	1	0	1/2	0	1/2	−2	2	1/2	−1/2	0
	2	1/2	−1/2	0	0	1/2	−1/2	1/8	1/2	1/8
$\pi/4$	0	0	3/2	−3/2	0	3/2	−3/2	3/4	−3/2	3/4
	1	1/4	−3/4	1	1/4	−3/4	1	0	1/2	−1/2
	2	1/4	0	−1/4	1/4	0	−1/4	0	1/4	0
$\pi/2$	0	0	3	−3	0	0	0	0	0	0
	1	1/2	−2	2	0	1/2	0	1/2	−1/2	0
	2	0	1/2	−1/2	1/2	−1/2	0	1/8	1/2	1/8

Table 3.8 Classification of the Second-Rank Irreducible Representations $\Gamma_\alpha^{(2)}$ and the Explicit Expressions of the Average One-Fold Probability Transform A_α of the Crystallographic Point Symmetry Groups of Pure Rotation G

Crystal system	G	$\Gamma_\alpha^{(2)}$	α	$A_\alpha = \chi_{\alpha E}^{-1} \sum_i p_i \chi_i$
Cubic	O	Γ_3, E	1	$p(E) - p(C_3)/2 + p(C_4^2)$
		Γ_5, F_2	2	$p(E) + p(C_2)/3 - p(C_4)/3 - p(C_4^2)/3$
	T	Γ_{23}', E	1	$p(E) + p(C_2) - p(C_3)/2$
		Γ_4', F	2	$p(E) - p(C_2)/3$
Tetragonal	D_4	Γ_{t1}, A_1	0	1
		Γ_{t3}, B_1	1	$p(E) + p(C_2) - p(C_4) + p(U_2) - p(U_2')$
		Γ_{t4}, B_2	2	$p(E) + p(C_2) - p(C_4) - p(U_2) + p(U_2')$
		Γ_{t5}, E	3	$p(E) - p(C_2)$
	C_4	Γ_{t1}', A	0	1
		${}^1\Gamma_{t2}'$, ${}^1B$	1	$p(E) + p(C_2) - p(C_4) - p(C_4^3)$
		${}^2\Gamma_{t2}'$, ${}^2B$	2	$p(E) + p(C_2) - p(C_4) - p(C_4^3)$
		Γ_{t34}', E	3	$p(E) - p(C_2)$
Hexagonal	D_6	Γ_{h1}, A	0	1
		Γ_{h5}, E_1	1	$p(E) - p(C_2) - p(C_3)/2 + p(C_6)/2$
		Γ_{h6}, E_2	2	$p(E) + p(C_2) - p(C_3)/2 - p(C_6)/2$
	C_6	Γ_{h1}', A	0	1
		Γ_{h34}', E_1	1	$p(E) - p(C_2) - p(C_3)/2 + p(C_6)/2$
		Γ_{h56}', E_2	2	$p(E) + p(C_2) - p(C_3)/2 - p(C_6)/2$
Trigonal	D_3	Γ_1^T, A	0	1
		${}^1\Gamma_3^T$, ${}^1E$	1	$p(E) - p(C_3)/2$
		${}^2\Gamma_3^T$, ${}^2E$	2	$p(E) - p(C_3)/2$
	C_3	$\Gamma_1'^T$, A	0	1
		${}^1\Gamma_2'^T$, ${}^1E$	1	$p(E) - p(C_3)/2$
		${}^1\Gamma_2'^T$, ${}^1E$	2	$p(E) - p(C_3)/2$
Orthorhombic	D_2	$2\Gamma_{r1}$, 2A	0	1
		Γ_{r2}, B_1	1	$p(E) - p({}^1C_2) - p({}^2C_2) + p({}^3C_2)$
		Γ_{t4}, B_2	2	$p(E) - p({}^1C_2) + p({}^2C_2) - p({}^3C_2)$
		Γ_{r3}, B_3	3	$p(E) + p({}^1C_2) - p({}^2C_2) - p({}^3C_2)$
Monoclinic	C_2	$3\Gamma_{m1}$, 3A	0	1
		${}^1\Gamma_{m1}$, ${}^1B$	1	$p(E) - p(C_2)$
		${}^2\Gamma_{m2}$, ${}^2B$	2	$p(E) - p(C_2)$

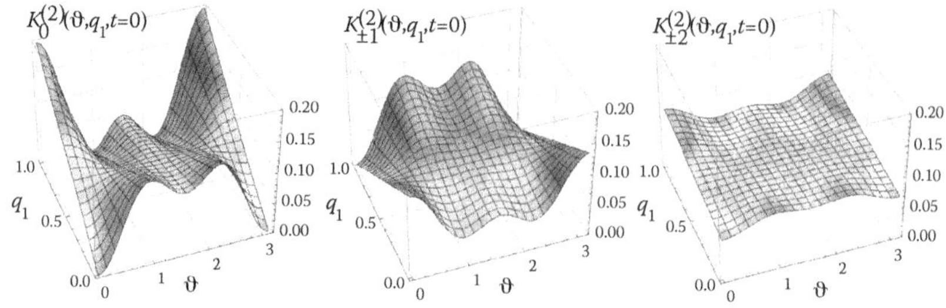

Figure 3.11 Graphs of the peak value ACFs $K_0^{(2)}(\vartheta,q_1,t=0)$, $K_{\pm1}^{(2)}(\vartheta,q_1,t=0)$, and $K_{\pm2}^{(2)}(\vartheta,q_1,t=0)$ adapted to the point symmetry groups O and T, drawn as functions of the polar angle ϑ of the CRF orientation in the LRF and the dynamical weight q_1. Fitted parameter $\phi = \pi/4$.

Figure 3.12, the peak values of the cubic system ACFs are drawn as functions of the polar angle ϑ and the dynamical weight of two-dimensional representation $q_1^{(2)} = q_1$, where the azimuthal angle ϕ equals respectively to $\phi = \pi/4$ and $\phi = 0$. We can see a straight-line parallel to the θ axis in these graphs that indicates the absence of anisotropy. This is the case when the dynamical weights are equal to the corresponding static ones. In accordance with Equation (2.3), they are equal to $q_1^{(2)} = q_1 = \frac{2}{5} = 0.4$ and $q_2^{(2)} = q_2 = \frac{3}{5} = 0.6$.

Examples of graphs displaying angular dependence of the peak values of the second-rank ACFs adapted to the groups of hexagonal-trigonal system are shown in Figure 3.13 and Figure 3.14. The weight of the hindered state symmetrized on the identical irreducible representation is taken as $q_0 = 0$ (Figure 3.13) and $q_0 = 0.2$ (Figure 3.14), and the azimuthal angle ϕ is arbitrary. The anisotropy of these ACFs never vanishes by varying the dynamical weights.

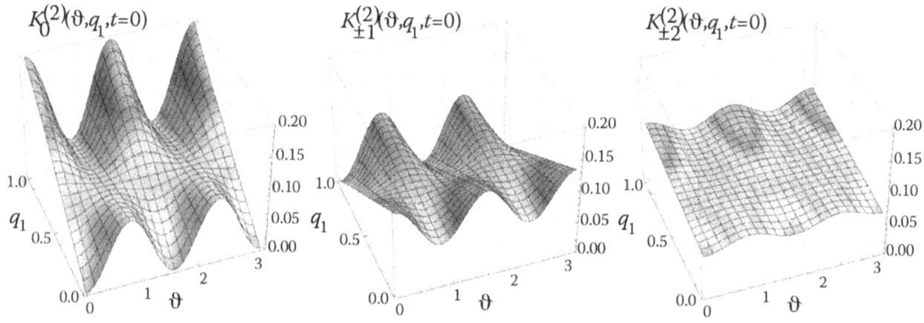

Figure 3.12 Graphs of the peak value ACFs $K_0^{(2)}(\vartheta,q_1,t=0)$, $K_{\pm1}^{(2)}(\vartheta,q_1,t=0)$, and $K_{\pm2}^{(2)}(\vartheta,q_1,t=0)$ adapted to the point symmetry groups O and T, drawn as functions of the polar angle ϑ of the CRF orientation in the LRF and the dynamical weight q_1. Fitted parameter $\phi = 0$.

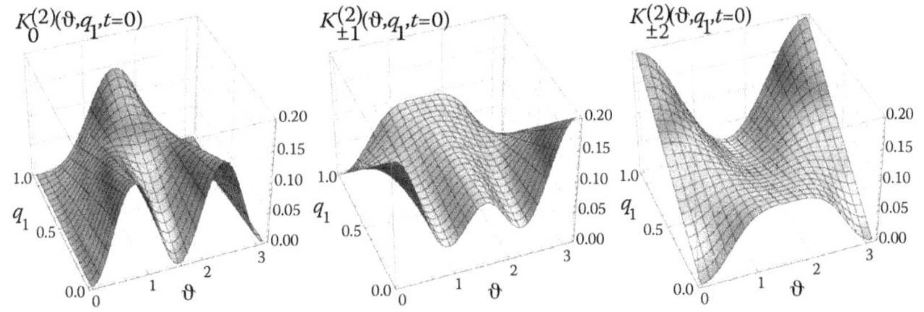

Figure 3.13 Graphs of the peak value ACFs $K_0^{(2)}(\vartheta, q_1, t=0)$, $K_{\pm1}^{(2)}(\vartheta, q_1, t=0)$, and $K_{\pm2}^{(2)}(\vartheta, q_1, t=0)$ adapted to the point symmetry groups of the hexagonal-trigonal system, drawn as functions of the polar angle ϑ of the CRF orientation in the LRF and the dynamical weight q_1. Fitted parameter $q_0 = 0$.

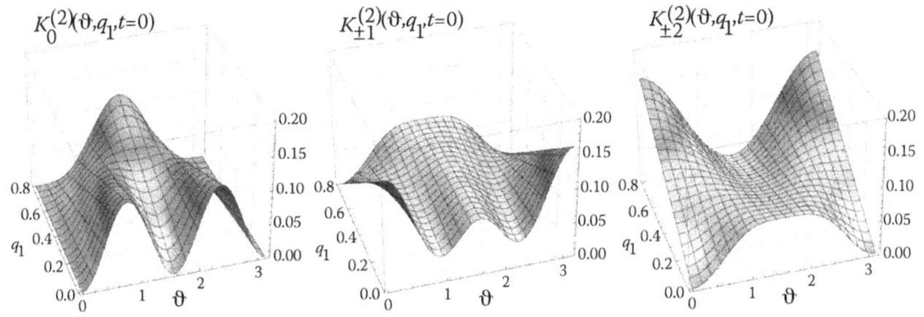

Figure 3.14 Graphs of the peak value ACFs $K_0^{(2)}(\vartheta, q_1, t=0)$, $K_{\pm1}^{(2)}(\vartheta, q_1, t=0)$, and $K_{\pm2}^{(2)}(\vartheta, q_1, t=0)$ adapted to the point symmetry groups of the hexagonal-trigonal system, drawn as functions of the polar angle ϑ of the CRF orientation in the LRF and the dynamical weight q_1 ($q_2 = 0.8 - q_1$). Fitted parameter $q_0 = 0.2$.

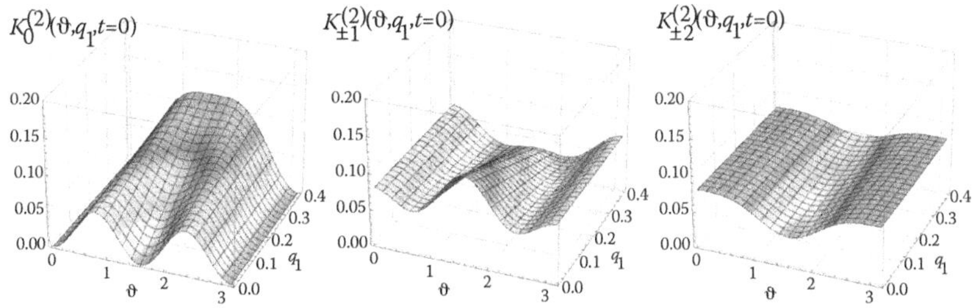

Figure 3.15 Graphs of the peak value ACFs $K_0^{(2)}(\vartheta, q_1, t=0)$, $K_{\pm1}^{(2)}(\vartheta, q_1, t=0)$, and $K_{\pm2}^{(2)}(\vartheta, q_1, t=0)$ adapted to the point symmetry groups of the tetragonal system, drawn as functions of the polar angle ϑ of the CRF orientation in the LRF and the dynamical weight q_1. The fitted parameters are $\phi = 0$, $q_0 = 0.2$, $q_3 = 0.4$, $q_2 = 0.4 - q_1$.

The peak values of the tetragonal system ACFs are drawn as functions of the polar angle ϑ and the dynamical weight of one two-dimensional representation $q_1^{(2)} = q_1$ in Figure 3.15. Other dynamical weights are fitted by $q_0 = 0.2$, $q_2 = 0.4 - q_1$, and $q_3 = 0.4$. The azimuthal angle is $\phi = 0$.

3.4 Discussion

- The symmetry-adapted ACFs given by Equations (3.4), (3.6) through (3.9), and (3.48) consist of the sum of the components symmetrized on the irreducible representations of crystallographic point groups and include the exponentially decreasing terms as well as those constants. The decreasing components are symmetrized on the non-identical, irreducible representations. The constant term of the ACF is symmetrized on the identical irreducible representations of the HMM point symmetry group. The total number of terms in an ACF of the rank ν does not exceed the dimension ($\mu_\nu = 2\nu + 1$) of the representation D_ν reducible on the given point group G. In single crystals, the number of terms can be extremely reduced by means of the particular choice of the crystal orientation.
- Unlike the previous HMM theories, the approach presented here takes into consideration the dual symmetry effects in the molecular quantity: the symmetry of the molecular motion and that of its site. The symmetrized correlation times τ_α account for the motional symmetry, whereas the dynamical weights of the hindered states $q_\alpha^{(\nu)}$ characterize the site symmetry. Two dynamical parameters τ_α and $q_\alpha^{(\nu)}$ allow us to obtain extended knowledge of the microscopic behavior of molecules from molecular spectra.
- With respect to Equation (2.29), the symmetrized correlation times τ_α relate to τ, the average time between two successive steps of the motion, and p_i, the probability of the molecular rearrangement by the symmetry elements forming *i*th class of the HMM group. According to Equation (3.8), the identity symmetry element presents the ordinary element of the stochastic process. It contributes to the correlation time τ_α and, hence, in the ACF [Equation (3.8)]. This result rectifies a mistake of former HMM models, where the identity symmetry element has been omitted from consideration.
- The dynamical weights $q_\alpha^{(\nu)}$ introduce the molecular site symmetry effect in the variation of the molecular quantity. According to Neumann's crystallographic principle, the anisotropy of the second-rank ACF in a cubic site manifests the crystal structure distortion. With respect to Equations (2.2) and (2.3), the second-rank dynamical weights corresponding to the ideal cubic sites are equal to $q_1 = 0.4$ and $q_2 = 0.6$. The ACF anisotropy does not arise in this case (Figure 3.11 and Figure 3.12). The deviation of the weights q_1 and q_2 from the

respective static values 0.4 and 0.6 allows one to estimate the type of symmetry distortion.

It was noted that the first-rank ACF adapted to a group of the cubic system is described by a single dynamical weight that is always equal to its static value. This ACF is isotropic. Hence it is impossible to determine the distortion of a local cubic symmetry by performing a spectroscopic experiment with a first-rank tensor molecular quantity.

chapter four

Dielectric and optical spectroscopy applications

Hindered molecular motion (HMM) study is commonly conducted by familiar spectroscopy techniques such as dielectric, infrared, Rayleigh, and Raman spectroscopy experimental procedures that make use of continuous wave technology [6–14]. This consists of randomly rotating permanent or induced electric dipoles, giving rise to lines of molecular spectra forced by the electric component of incident radiation. In the cited experimental spectroscopy techniques, the shape of spectral lines involves basic knowledge of the peculiarities of HMM.

4.1 *Frequency domain dielectric spectroscopy*

Using dielectric spectroscopy to study problems in condensed molecular systems has several advantages. These include the detailed nature of the information that can be obtained on structure and dynamic properties, the certainty of interpretation that may be given to the data, and the exceedingly wide range of types of sample and experimental circumstances. Dielectric properties reflect the structure of the matter and the dynamics of molecules and phase condition, as well as the effects of temperature and pressure to them. Small atomic formations and molecules, which exhibit a hindered motion in some crystals and liquids, give rise to the energy loss of an alternating electric field at frequencies lower than 10^{12} Hz. The fundamentals of dielectric properties of molecular substances deal with the concept of complex permittivity ε^* [6–9, 13] presented by

$$\varepsilon^* = \varepsilon' - i\varepsilon'' \tag{4.1}$$

where ε' and ε'' are the real and imaginary parts of ε^*. Usually, one examines the dielectric permittivities ε' and ε'' as functions of the frequency and the strength of the external alternating electric field, temperature, and pressure. Since the days of Debye, the time-dependent angular autocorrelation functions (ACF) of the macroscopic electric polarization have been applied to calculate the complex permittivity ε^* in molecular liquids. The explicit form of the ACF includes its dependence on the HMM model. The ACF depends additionally on the crystal orientation with respect to the electric field strength in molecular crystals.

HMM forces molecules into a random change of static and induced electrical dipole moments. Owing to a small time lag, the induced moments give an appreciable contribution to the permittivity ε^* at high frequencies, falling to the region of infrared radiation. At low frequencies, the polarization of a polar dielectric is commonly due to the orientation ordering of polar molecules.

In the case of the relaxation polarization of a dielectric placed in a feeble alternating electric field, we can calculate the complex permittivity of the matter independently of the HMM model by using the ordinary expression [8, 9]:

$$\frac{\varepsilon^*(\omega)-\varepsilon_\infty}{\varepsilon_s-\varepsilon_\infty}=1-\mathrm{i}\,\omega\int_0^\infty \Phi(t)\exp(-\mathrm{i}\,\omega t)\,\mathrm{d}t \tag{4.2}$$

where ε_s and ε_∞ are, respectively, the static dielectric permittivity and the permittivity measured at very high frequencies ω of the external electric field, and $\Phi(t)$ is the normalized ACF of the macroscopic polarization of matter.

For independent dipoles, $\Phi(t)$ is the normalized ACF of the "cosine" of the polar angle θ that relates the direction of the dipole moment to the electric field strength. In terms of spherical tensors, the function $\Phi(t)$ is equal to the ACF $K_0^{(1)}(t)$ of the zero component of the first-rank normalized spherical harmonic $Y_0^{(1)}(\theta,\varphi)$ up to the factor 4π:

$$\Phi(t)=\frac{\langle\cos\theta_0\cos\theta\rangle}{\langle\cos^2\theta_0\rangle}=\frac{\left\langle Y_0^{(1)}(\theta_0,\varphi_0)Y_0^{(1)}(\theta_0,\varphi_0)^*\right\rangle}{\left\langle\left|Y_0^{(1)}(\theta_0,\varphi_0)\right|^2\right\rangle}=4\pi K_0^{(1)}(t) \tag{4.3}$$

Here, angular brackets include time averaging as well as statistical averaging over all molecules of the substance. With the help of Equation (4.3), the basic Equation (4.2) takes the form

$$\frac{\varepsilon^*(\omega)-\varepsilon_\infty}{\varepsilon_s-\varepsilon_\infty}=1-4\pi\mathrm{i}\,\omega\int_0^\infty K_0^{(1)}(t)\exp(-\mathrm{i}\,\omega t)\,\mathrm{d}t \tag{4.4}$$

We can derive the explicit expression of the complex dielectric permittivity $\varepsilon^*(\omega)$ if the analytical form of ACF $K_0^{(1)}(t)$ is known. We use the single-crystalline expression $K_0^{(1)}(t)=K_0^{(1)}(\phi,\vartheta,t)$ derived within the framework of the extended angular jump (EAJ) model by

$$K_0^{(1)}(\phi,\vartheta,t)=\frac{3}{4\pi}\sum_{\alpha}\sum_{l=0,1}q_\alpha^{(1)}\,a_{\alpha l0}^{(1)}(\phi)\cos^{2l}\vartheta\exp\left(-\frac{t}{\tau_\alpha}\right) \tag{4.5}$$

where ϕ and ϑ are the azimuthal and polar angles of the vector of electric field strength determined with respect to the main axis of the crystallographic reference frame (CRF). By substituting Equation (4.5) into Equation (4.4), performing some transforms, and separating the real part from the imaginary part, we can derive the reduced dielectric constant η' and the reduced loss factor η'' as functions of ϕ, ϑ, and ω in single-crystalline samples. They are equal respectively to [69, 70]

$$\eta'(\phi,\vartheta,\omega)=\frac{\varepsilon'(\phi,\vartheta,\omega)-\varepsilon_\infty}{\varepsilon_0-\varepsilon_\infty}=1-3\sum_{\alpha=1}^{3}\sum_{l=0,1}q_\alpha^{(1)}a_{\alpha l0}^{(1)}(\phi)\cos^{2l}\vartheta\frac{\omega^2}{\omega^2+\omega_\alpha^2} \quad (4.6)$$

and

$$\eta''(\phi,\vartheta,\omega)=\frac{\varepsilon''(\phi,\vartheta,\omega)}{\varepsilon_0-\varepsilon_\infty}=3\sum_{\alpha=1}^{3}\sum_{l=0,1}q_\alpha^{(1)}a_{\alpha l0}^{(1)}(\phi)\cos^{2l}\vartheta\frac{\omega\omega_\alpha}{\omega^2+\omega_\alpha^2} \quad (4.7)$$

In powder samples, they reduce to

$$\eta'(\omega)=\frac{\varepsilon'(\omega)-\varepsilon_\infty}{\varepsilon_0-\varepsilon_\infty}=1-\sum_{\alpha=1}^{3}q_\alpha^{(1)}\frac{\omega^2}{\omega^2+\omega_\alpha^2} \quad (4.8)$$

and

$$\eta''(\omega)=\frac{\varepsilon''(\omega)}{\varepsilon_0-\varepsilon_\infty}=\sum_{\alpha=1}^{3}q_\alpha^{(1)}\frac{\omega\,\omega_\alpha}{\omega^2+\omega_\alpha^2} \quad (4.9)$$

where the variable ω_α, the symmetrized frequency of the dipole moment correlation, is used instead of τ_α, the symmetrized correlation time. The relation between the variables τ_α and ω_α is

$$\omega_\alpha=1/\tau_\alpha \quad (4.10)$$

Consequently,

$$\omega_\alpha=\omega_{\alpha 0}A_\alpha=\omega_{\alpha 0}\left(1-\chi_{\alpha E}^{-1}\sum_i\chi_{\alpha i}p_i\right) \quad (4.11)$$

where A_α is the average one-fold probability transform and $\omega_{\alpha 0}=1/\tau_{\alpha 0}$ is the average frequency of attempts to overcome the barrier of the motion. For the pure rotation crystallographic point symmetry groups, most important in applications, the explicit expressions of the factors A_α and $a_{\alpha l0}^{(1)}(\phi)$ are presented in Tables 3.3 and 3.4.

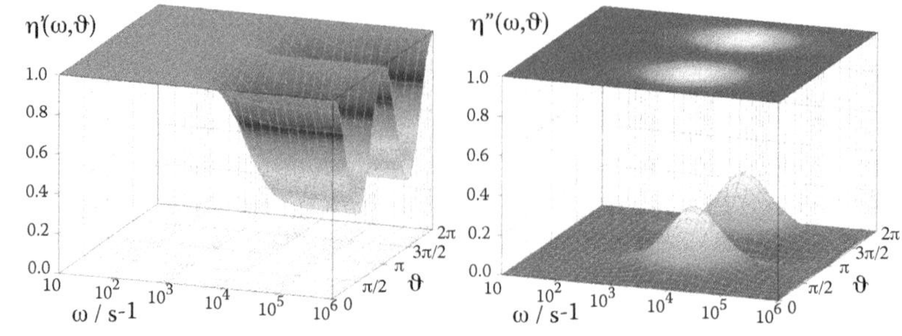

Figure 4.1 Theoretical graphs of the reduced dielectric constant η′ and the loss factor η″ drawn as functions of ω, the frequency of alternating electric field, and ϑ, the polar angle of the single-crystal orientation, for the HMM point symmetry groups C_n ($n = 3, 4, 6$). The fitted parameters are equal to $q_0^{(1)} = 1/3$, $q_1^{(1)} = 2/3$, and $\omega_1 = 10^4\ s^{-1}$.

To throw more light upon the anisotropic properties of the reduced dielectric constant and the reduced loss factor, some surface plot graphs of the spectra predicted by Equations (4.6) and (4.7) are shown in Figure 4.1 through Figure 4.6. They are drawn for the HMM adapted to the crystallographic point symmetry groups of pure rotation C_n ($n = 3, 4, 6$), D_n ($n = 3, 4, 6$), C_2, and D_2. We can see that the reduced dielectric constant does not decrease to zero for some crystal orientations if the HMM is adapted to the groups C_n, D_2, and C_2. There is no anisotropic behavior of quantities η′ and η″, whereas the HMM is adapted to a point symmetry group of cubic system.

Let us examine the polycrystalline results. The examples of the theoretical frequency dependences of η′(ω) and η″(ω), the reduced dielectric constant and the reduced loss factor, respectively, are shown in

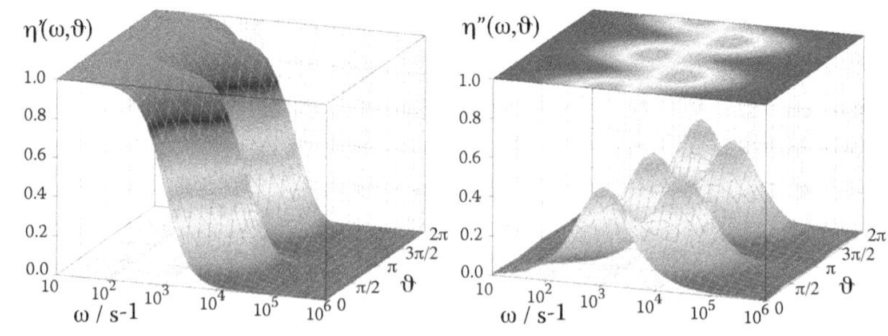

Figure 4.2 Theoretical graphs of the reduced dielectric constant η′ and the loss factor η″ drawn as functions of ω, the frequency of alternating electric field, and ϑ, the polar angle of single-crystal orientation, for the HMM point symmetry groups D_n ($n = 3, 4, 6$). The fitted parameters are equal to $q_1^{(1)} = 2/3$, $q_2^{(1)} = 1/3$, $\omega_1 = 10^4\ s^{-1}$, and $\omega_2 = 10^3\ s^{-1}$.

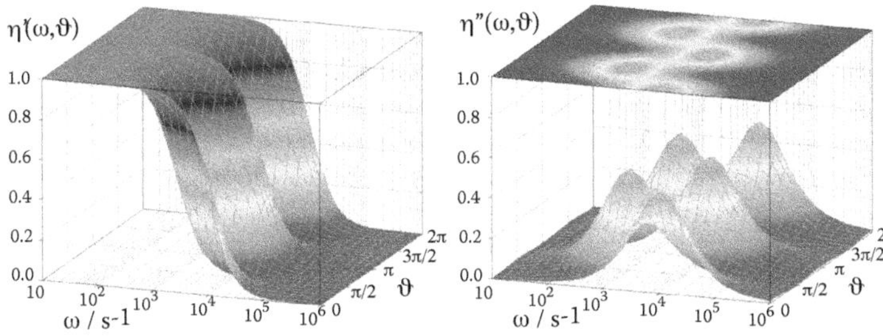

Figure 4.3 Theoretical graphs of the reduced dielectric constant η′ and the reduced loss factor η″ drawn as functions of ω, the frequency of alternating electric field, and ϑ, the polar angle of single-crystal orientation, for the HMM point symmetry groups D_n ($n = 3, 4, 6$). The fitted parameters are equal to $q_1^{(1)} = 2/3$, $q_2^{(1)} = 1/3$, $\omega_1 = 10^3\ s^{-1}$, and $\omega_2 = 10^4\ s^{-1}$.

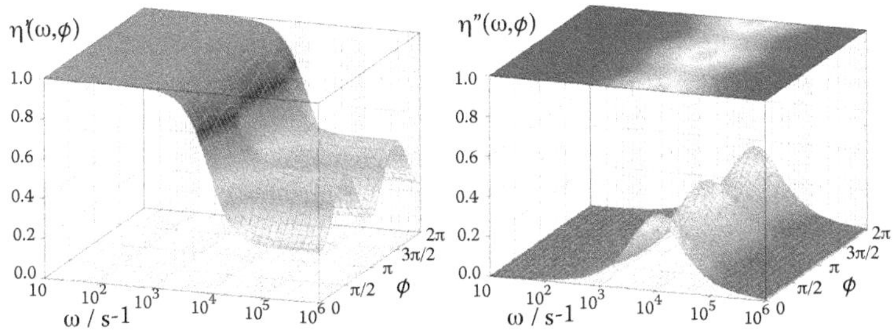

Figure 4.4 Theoretical graphs of the reduced dielectric constant η′ and the reduced loss factor η″ drawn as functions of ω, the frequency of alternating electric field, and ϕ, the azimuthal angle of single crystal orientation, for the HMM point symmetry group C_2. The fitted parameters equal to $\vartheta = 90^0$, $q_0^{(1)} = 0.2$, $q_1^{(1)} = 0.5$, $q_2^{(1)} = 0.3$, and $\omega_1 = \omega_2 = 10^4\ s^{-1}$.

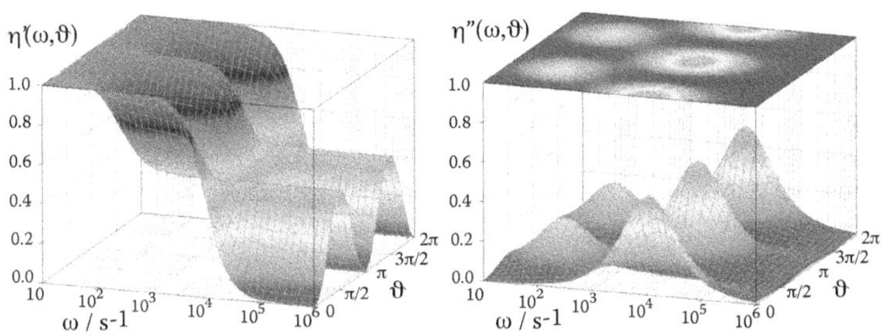

Figure 4.5 Theoretical graphs of the reduced dielectric constant η′ and the reduced loss factor η″ drawn as functions of ω, the frequency of alternating electric field, and ϑ, the polar angle of single crystal orientation, for the HMM point symmetry group D_2. The fitted parameters equal to $\phi = 0$, $q_1^{(1)} = q_2^{(1)} = q_3^{(1)} = 1/3$, $\omega_1 = 10^2\ s^{-1}$, $\omega_2 = 10^3\ s^{-1}$, and $\omega_3 = 10^4\ s^{-1}$.

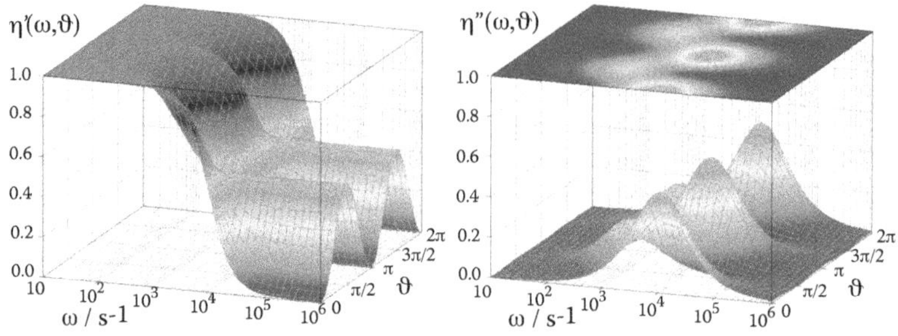

Figure 4.6 Theoretical graphs of the reduced dielectric constant η′ and the reduced loss factor η″ drawn as functions of ω, the frequency of alternating electric field, and ϑ, the polar angle of single crystal orientation, for the HMM point symmetry group D_2. The fitted parameters equal to $\phi = 90°$, $q_1^{(1)} = q_2^{(1)} = q_3^{(1)} = 1/3$, $\omega_1 = 10^2\ s^{-1}$, $\omega_2 = 10^3\ s^{-1}$, and $\omega_3 = 10^4\ s^{-1}$.

Figure 4.7(a) through 4.7(f). They are drawn for dipoles exhibiting random rotations adapted to the symmetry groups: O or T, Figure 4.7(a); D_n ($n = 3, 4, 6$), Figure 4.7(b) through 4.7(d); D_2, Figure 4.7(e); and C_n ($n = 3, 4, 6$), Figure 4.7(f). The curves displayed in Figure 4.7(a) through 4.7(e) have the same shape as prescribed by the Debye relaxation polarization theory for isotropic rotational diffusion of molecules with one diffusion constant [Figure 4.7(a)], the anisotropic rotational diffusion with two diffusion constants [Figure 4.7(b) through 4.7(d)], and anisotropic rotational diffusion with three diffusion constants [Figure 4.7(e)].

However, the curves shown in Figure 4.7(f) are not prescribed by Debye's theory: The reduced constant does not fall to zero at high frequencies and the maximal value of the reduced loss factor, which takes place at $\omega = \omega_1$, is not equal to 1/2. This divergence from Debye's theory is due to accounting for the HMM symmetrized on the identical representation in the EAJ model framework. Such dielectric property allows us to distinguish the types of HMM models in a system of polar molecules exhibiting axial motion: the EAJ model from the rotational diffusion (RD) model. The corresponding Cole–Cole diagrams are shown in Figure 4.8. The curves displayed in Figure 4.7 and Figure 4.8 are frequently observed in condensed molecular media [6–9, 13, 71].

4.2 *Shape of polarized infrared absorption spectroscopy lines*

The physical methods based on examining the frequencies and intensities of molecular spectral lines of emission, absorption, and reflection of infrared radiation are largely applied to studying the dynamics and structure of molecules. An infrared spectrum arises from electrical dipole transitions

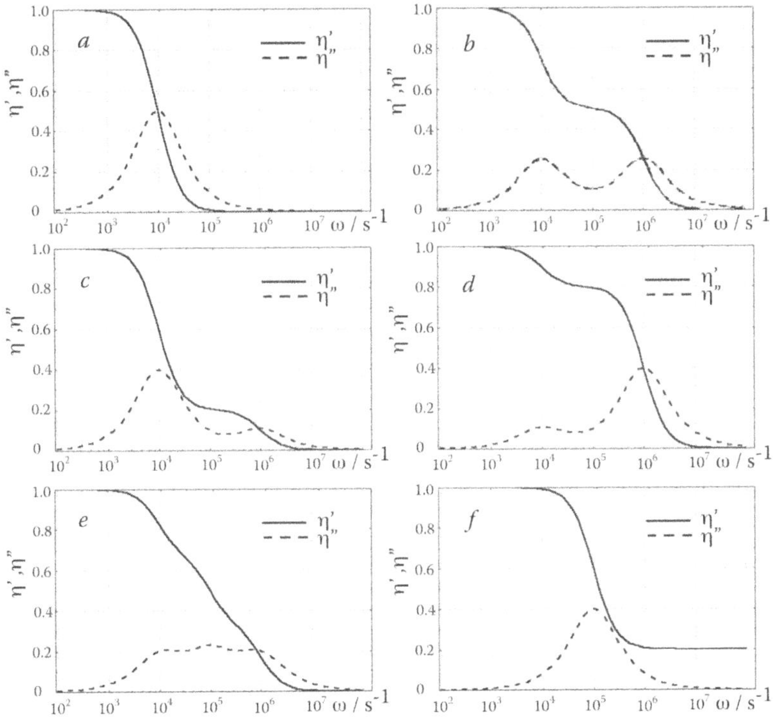

Figure 4.7 Graphs of the reduced dielectric constant η′ and the reduced loss factor η″ drawn as functions of ω, the frequency of the alternating electric field in molecular polycrystals, for the HMM adapted to the point symmetry groups of pure rotation: (a) O or T with $\omega_1 = 10^4$ s^{-1}; (b) D_n ($n = 3, 4, 6$) with $q_1^{(1)} = q_2^{(1)} = 0.5$, $\omega_1 = 10^4$ s^{-1}, and $\omega_2 = 10^6$ s^{-1}; (c) D_n ($n = 3, 4, 6$) with $q_1^{(1)} = 0.8$, $q_2^{(1)} = 0.2$, $\omega_1 = 10^4$ s^{-1}, and $\omega_2 = 10^6$ s^{-1}; (d) D_n ($n = 3, 4, 6$) with $q_1^{(1)} = 0.2$, $q_2^{(1)} = 0.8$, $\omega_1 = 10^4$ s^{-1}, and $\omega_2 = 10^6$ s^{-1}; (e) D_2 with $q_1^{(1)} = q_2^{(1)} = q_3^{(1)} = 1/3$, $\omega_1 = 10^4$ s^{-1}, $\omega_2 = 10^5$ s^{-1}, and $\omega_3 = 10^6$ s^{-1}; (f) C_n ($n = 3, 4, 6$) with $q_0^{(1)} = 0.2$, $q_1^{(1)} = 0.8$, and $\omega_1 = 10^5$ s^{-1} and C_2 with $q_0^{(1)} = 0.2$, $q_1^{(1)} + q_2^{(1)} = 0.8$, and $\omega_1 = 10^5$ s^{-1}.

in a system of molecular vibration and rotational energy levels. At ambient temperatures, the molecules occupy mainly the ground electronic, vibrational, and rotational states in liquids and solids. Therefore, there is a small emission of electromagnetic radiation and the study of the infrared spectrum is available under absorption or reflection of radiation by the matter.

A molecule composed of two or more atoms possesses translational, vibrational, and rotational degrees of freedom. The excitation of a certain vibration induces an alternating electric dipole moment of the molecule. Absorption of the electromagnetic energy by molecular dipoles takes place at frequencies close to frequencies of the dipole moment alteration.

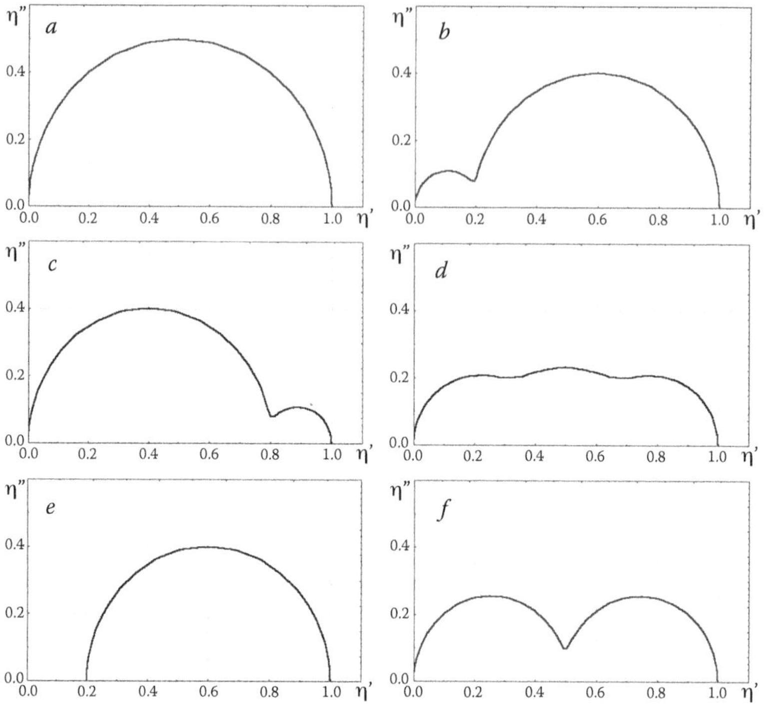

Figure 4.8 Cole-Cole diagrams drawn in accordance with the spectra shown in Figure 4.7.

The HMM displays a continuous spectrum and produces the incoherent modulation of the oscillation. The temperature dependence of the HMM average frequency follows Arrhenius's law. Therefore, the broadening produced by HMM can be studied irrespective of all other kinds of broadening.

The generality of the theory of infrared line broadening due to molecular motion is given elsewhere [10–13]. In accordance with Gordon [11, 26], we calculate the expression for the normalized spectral density function (SDF) of infrared absorption, the relative intensity of the infrared vibration band, as a function of frequency by applying the Fourier transform to the molecular dipole moment correlation function. Within the limits of one-particle approximation, the base expression for this SDF is:

$$J(\omega) = \frac{3}{2\pi} \int_{-\infty}^{\infty} \langle(\boldsymbol{\varepsilon\mu}(0))(\boldsymbol{\varepsilon\mu}(t))\rangle e^{-i(\omega-\omega')t} dt \tag{4.12}$$

where ω is the frequency of incident light, ω' is the vibrational frequency of an electrical dipole moment, $\boldsymbol{\varepsilon}$ is the unit vector of alternating electric field strength, $\boldsymbol{\mu}(0)$ and $\boldsymbol{\mu}(t)$ are the unit vectors of the dipole moment fixed at the initial instant $t = 0$ and the arbitrary time t, respectively. Angular brackets denote averaging over all molecules of the substance and the time. The quantity $\langle(\boldsymbol{\varepsilon\mu}(0))(\boldsymbol{\varepsilon\mu}(t))\rangle$ is the ACF of the cosine of the angle that the electrical dipole moment makes with respect to the field strength. We shall express the required ACF by the first-rank ACF $K_0^{(1)}(t)$

$$\begin{aligned}\langle(\boldsymbol{\varepsilon\mu}(0))(\boldsymbol{\varepsilon\mu}(t))\rangle &= \langle\cos(\theta_0)\cos(\theta)\rangle = \frac{4\pi}{3}\left\langle Y_0^{(1)}(g_0)Y_0^{(1)}(g)\right\rangle \\ &= \frac{4\pi}{3}\oiint Y_0^{(1)}(g_0)Y_0^{(1)}(g)\, W(g_0,t,g)\mathrm{d}g_0\mathrm{d}g = \frac{4\pi}{3}K_0^{(1)}(t)\end{aligned} \tag{4.13}$$

By substituting Equation (4.13) into Equation (4.12), the expression of the SDF takes the form:

$$J(\omega) = J_0^{(1)}(\omega) = 2\int_{-\infty}^{\infty} K_0^{(1)}(t)e^{-i(\omega-\omega')t}\mathrm{d}t \tag{4.14}$$

By using the analytical expression of the first-rank ACF given by Equation (4.5) and integrating Equation (4.14) under the condition that only the imaginary part is responsible for infrared absorption, we obtain the required expression for the normalized SDF

$$J_0^{(1)}(\phi,\vartheta,\omega) = \frac{3}{2\pi}\sum_{\alpha}\sum_{l=0,1} q_\alpha^{(1)} a_{\alpha l0}^{(1)}(\phi)\cos^{2l}\vartheta \frac{\omega_\alpha}{(\omega-\omega')^2+\omega_\alpha^2} \tag{4.15}$$

The explicit expressions of the factors $a_{\alpha l0}^{(1)}(\phi)$ are presented in Table 3.4. Equation (4.11) gives the correlation frequencies ω_α. In powders, Equation (4.15) reduces to

$$J_0^{(1)}(\omega) = \frac{1}{2\pi}\sum_{\alpha} q_\alpha^{(1)} \frac{\omega_\alpha}{(\omega-\omega')^2+\omega_\alpha^2} \tag{4.16}$$

The angular dependence of the SDFs predicted by Equation (4.15) is the same as for the reduced dielectric loss factors η'' expressed by Equation (4.7). Therefore, the graphs of the ACFs displayed in Figure 3.5 and Figure 3.7 through Figure 3.10 also present graphs of the relative peak intensities of the infrared absorption lines as functions of the angles of crystal orientation (ϑ, ϕ) and the dynamical weights $q_\alpha^{(1)}$. The frequency dependence of the SDFs gives the absorption lines whose shape has Lorentzian form.

4.3 Line shape of Rayleigh and Raman light scattering

The molecular spectra of Rayleigh and Raman scattering of light arise owing to the interaction of the incident electromagnetic radiation with the induced electrical dipole moments and correspond to certain types of intra- and intermolecular vibrations [10–15]. The HMM results in a homogeneous broadening of Rayleigh and Raman scattering lines in the same way as its effect in the infrared spectral lines. In the framework of classical physics, the vector of induced electrical dipole moment **d** links to the vector of electrical field intensity **E** by the expression

$$\mathbf{d} = \alpha\,\mathbf{E} \tag{4.17}$$

where α is the polarizability tensor, a function of the nuclear coordinates, and hence it depends on the frequencies of molecular vibration. To simplify the treatment, but without loss of generality, we assume that a molecule possesses only one vibrational degree of freedom. We shall extend the vector **d** by adding the frequency dependence of both **E** and α into Equation (4.17). The variation of the polarizability with the vibration of the molecule can be expressed by expanding each component of the polarizability tensor α in a Taylor series of Q, the normal coordinate—the vibration coordinate near the equilibrium configuration of the molecule. The terms that involve powers of Q higher than the first power will be neglected here. Thus, the Taylor series of α_{ij} can be written in the following manner:

$$\alpha_{ij} = \alpha_{ij0} + \frac{\partial \alpha_{ij}}{\partial Q} Q + \cdots = \alpha_{ij0} + \alpha'_{ij} Q + \cdots \tag{4.18}$$

where α_{ij0} is the value of α_{ij} determined close to the equilibrium configuration and $\alpha'_{ij} = \frac{\partial \alpha_{ij}}{\partial Q}$ is the component of a new tensor α'. The tensor α' is called the derived polarizability tensor, because all its components are the polarizability derivatives with respect to the normal coordinate Q.

Equation (4.18) is valid for all tensor components. Therefore, we can generalize it by

$$\alpha = \alpha_0 + \alpha' Q \tag{4.19}$$

where Q is a scalar quantity that multiplies all components α'. Assuming that a simple harmonic motion at frequency ω' associates with the molecule, the time dependence of Q can be written as

$$Q = Q_m \cos(\omega' t + \delta) \tag{4.20}$$

The expression for the monochromatic electric field strength is

$$\mathbf{E} = \mathbf{E}_m \cos \omega t \tag{4.21}$$

In Equations (4.20) and (4.21), Q_m and $\mathbf{E}_m$ are the amplitude values of Q and $\mathbf{E}$.

Substituting the values α and $\mathbf{E}$ given by Equations (4.19) through (4.21) into Equation (4.17) and performing some trigonometric transforms, we can present the expression for the frequency-dependent electric dipole moment as

$$\mathbf{d} = \mathbf{d}(\omega) + \mathbf{d}(\omega - \omega') + \mathbf{d}(\omega + \omega') \tag{4.22}$$

where the following notations

$$\left.\begin{aligned}
&\mathbf{d}(\omega) = \alpha_0 \mathbf{E}_m Q_m \cos \omega t = \alpha_{Ray} \mathbf{E}_m \cos \omega t, \\
&\mathbf{d}(\omega - \omega') = \frac{1}{2} \alpha' \mathbf{E}_m Q_m \cos\left[(\omega - \omega')t - \delta\right] = \alpha_{Ram} \mathbf{E}_m \cos\left[(\omega - \omega')t - \delta\right], \\
&\text{and} \\
&\mathbf{d}(\omega + \omega') = \frac{1}{2} \alpha' \mathbf{E}_m Q_m \cos\left[(\omega + \omega')t + \delta\right] = \alpha_{Ram} \mathbf{E}_m \cos\left[(\omega + \omega')t + \delta\right]
\end{aligned}\right\} \tag{4.23}$$

are used. There are two polarizability tensors in Equation (4.23). The tensor $\alpha_{Ray} = \alpha_0 Q_m$ is an important quantity in the theory of Rayleigh scattering. It is not equal to zero. Indeed, all molecules are capable of being polarized. Therefore, Rayleigh scattering is excited by default. The tensor $\alpha_{Ram} = (1/2)\alpha' Q_m$, important in the Raman effect, is often equal to zero. The Raman polarizability tensor does not equal zero if at least one derivative of the molecular polarizability component exists.

This relatively simple classical treatment provides us with a useful qualitative picture of the mechanisms of Rayleigh and Raman scattering. Rayleigh scattering arises from the electric dipole oscillated at ω, the frequency of the incident radiation whose electric component induces this dipole. Raman scattering appears from the electric dipoles oscillated at frequencies $\omega \pm \omega'$, where ω' is the frequency of the molecular vibration. The electrons, whose rearrangement imposes a harmonic variation of the polarizability due to nuclear motion, provide the necessary coupling between the nuclear motion and the electric field.

We describe the line shape of Rayleigh and Raman scattering in molecular crystals by using the EAJ model, chosen as the HMM model, under the following assumptions on the experimental procedure. Monochromatic

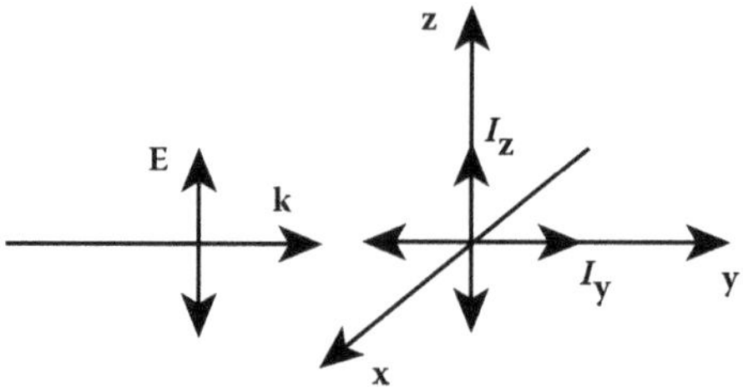

Figure 4.9 Schema of observation of the molecular scattering spectrum of light.

plane-polarized light $\mathbf{E}(0,0,E)$ falls to the sample along the y axis of the laboratory reference frame (LRF). In the direction x, the polarized ($I_z = I_{||}$) and depolarized ($I_y = I_{\perp}$) components of intensity of the scattered light are observed (Figure 4.9). Due to the conventional similarity of each formula in Equation (4.23), which expresses the induced dipole vectors active in both Rayleigh and Raman scattering, further theoretical treatment of the line shapes will not distinguish the two types of scattering.

The expressions of the relative intensities of the lines I_y and I_z can be derived by the Fourier transformation of the autocorrelation functions K_y and K_z of the respective components d_y and d_z of the induced electrical dipole moment **d**. The vector $\mathbf{d}(d_x, d_y, d_z)$ relates to the light vector **E** $(0,0,E)$ by means of the second-rank polarizability tensor α_{ij} in the LRF as

$$\begin{pmatrix} d_x \\ d_y \\ d_z \end{pmatrix} = \begin{pmatrix} \alpha_{xx} & \alpha_{xy} & \alpha_{xz} \\ \alpha_{yx} & \alpha_{yy} & \alpha_{yz} \\ \alpha_{zx} & \alpha_{xy} & \alpha_{zz} \end{pmatrix} \begin{pmatrix} 0 \\ 0 \\ E \end{pmatrix} \tag{4.24}$$

Consequently, the vector **d** is expressed by

$$\mathbf{d} = d_x\mathbf{i} + d_y\mathbf{j} + d_z\mathbf{k} = \alpha_{xz}E\mathbf{i} + \alpha_{yz}E\mathbf{j} + \alpha_{zz}E\mathbf{k} \tag{4.25}$$

From the last relation, we can write out the rating equations for the components d_y and d_z desired with respect to the schema of observation shown in Figure 4.9:

$$d_y = \alpha_{yz}E \quad \text{and} \quad d_z = \alpha_{zz}E \tag{4.26}$$

The polarizability tensor decomposes in two tensors, $\alpha_{ij}^{\mathrm{I}} = (^1/_3)\alpha_{ij}$ and $\alpha_{ij}^{\mathrm{II}} = \alpha_{ij} - \alpha_{ij}^{\mathrm{I}}\,\delta_{ij}$, the anisotropic and isotropic parts, respectively. The anisotropic part α_{ij}^{I} of the tensor α_{ij} depends on the molecular orientation and it is a symmetrical tensor with zero trace—that is, $\alpha_{ij}^{\mathrm{I}} = \alpha_{ji}^{\mathrm{I}}$ and $\Sigma_i\,\alpha_{ii} = 0$. The isotropic part $\alpha_{ij}^{\mathrm{II}}$ of the tensor α_{ij} contributes to the scattering due to only translational motion of the molecules and it can be neglected in the molecular crystals. Therefore, it will be omitted in further consideration. In order to simplify mathematical treatment of the light intensities by scattering, we shall transform the polarizability tensor given in the Cartesian basis α_{ij} to the polarizability tensor determined in the spherical one $\alpha_m^{(2)}$ by:

$$\begin{aligned} \alpha_0^{(2)} &= \sqrt{3/2}\ \alpha_{zz} \\ \alpha_{-1}^{(2)} &= (\alpha_{xz} - i\alpha_{yz}) \\ \alpha_1^{(2)} &= -(\alpha_{xz} + i\alpha_{yz}) \\ \alpha_{-2}^{(2)} &= \frac{1}{2}(\alpha_{xx} - \alpha_{yy} - 2i\alpha_{xy}) \\ \alpha_2^{(2)} &= \frac{1}{2}(\alpha_{xx} - \alpha_{yy} + 2i\alpha_{xy}) \end{aligned} \tag{4.27}$$

In terms of spherical tensors, Equation (4.24) can be rewritten in the form:

$$d_y = \frac{i}{2}\left(\alpha_1^{(2)} + \alpha_{-1}^{(2)}\right) \quad \text{and} \quad d_z = \sqrt{2/3}\,\alpha_0^{(2)}E \tag{4.28}$$

where $\alpha_m^{(2)}$ (m = –1, 0, 1) are the components of the spherical tensor created from the anisotropic part of the tensor α_{ij}. It should be noted that the components $\alpha_2^{(2)}$ and $\alpha_{-2}^{(2)}$ obtained from the symmetrical part of the tensor α_{ij} are equal to zero. The spherical tensor of the polarizability (or its derivative) of a symmetrical molecule determined in the reference frame of its principal axes of symmetry—that is, in the molecular reference frame (MRF)—has a unique nonzero element $\alpha_0^{(2)}$, which is a molecular constant. The spherical tensors transform under rotation with the help of Wigner matrices as the spherical harmonics. Equation (4.28) for quantities d_y and d_z can be rewritten as

$$d_y = \frac{i}{2}\alpha_0^{(2)}E\left[D_{01}^{(2)}(\Omega) + D_{0-1}^{(2)}(\Omega)\right] = i\sqrt{\pi/5}\ \alpha_0^{(2)}E\left[Y_1^{(2)}(\vartheta,\phi) + Y_{-1}^{(2)}(\vartheta,\phi)\right] \tag{4.29}$$

and

$$d_z = \sqrt{2/3}\ \alpha_0^{(2)}E\,D_{00}^{(2)}(\Omega) = -\sqrt{8\pi/15}\ \alpha_0^{(2)}E\,Y_0^{(2)}(\vartheta,\phi) \tag{4.30}$$

where $D_{01}^{(2)}(\Omega)$, $D_{0-1}^{(2)}(\Omega)$, and $D_{00}^{(2)}(\Omega)$ are the second-rank Wigner functions; $Y_1^{(2)}(\vartheta,\phi)$, $Y_{-1}^{(2)}(\vartheta,\phi)$, and $Y_0^{(2)}(\vartheta,\phi)$ are the second-rank normalized spherical harmonics (Table 3.2); Ω is the symbol for Euler angles $\Omega \equiv (\phi, \vartheta, \zeta)$, and (ϕ, ϑ) are the polar angles of the axis of molecular rotation symmetry in the LRF.

In the course of motion, angles ϕ and ϑ as well as components d_y and d_z show a random dependence on time. Therefore, the time dependence of the ACFs $K_y(t)$ and $K_z(t)$ is completely determined by the time dependence of the second-rank ACFs $K_1^{(2)}(t)$ and $K_0^{(2)}(t)$:

$$K_y(t) = \langle d_y(0) d_y(t)^* \rangle = \frac{2\pi}{5}\left[\alpha_0^{(2)} E\right]^2 K_1^{(2)}(t) \tag{4.31}$$

and

$$K_z(t) = \left\langle d_z(0) d_z(t)^* \right\rangle = \frac{8\pi}{15}\left[\alpha_0^{(2)} E\right]^2 K_0^{(2)}(t) \tag{4.32}$$

Applying Fourier transform to Equations (4.31) and (4.32), we can obtain the following expressions for $I_y = I_\perp$ and $I_z = I_{||}$, the relative intensities of the depolarized and polarized lines of the light scattering:

$$I_y(\Delta\omega) = I_\perp(\Delta\omega) = \frac{2\pi}{5}\left[\alpha_0^{(2)} E\right]^2 J_1^{(2)}(\Delta\omega) \tag{4.33}$$

and

$$I_z(\Delta\omega) = I_{||}(\Delta\omega) = \frac{8\pi}{15}\left[\alpha_0^{(2)} E\right]^2 J_0^{(2)}(\Delta\omega) \tag{4.34}$$

where $\Delta\omega$ is the frequency shift of Rayleigh scattering light from the frequency of the incident light. In the case of Raman scattering, $\Delta\omega$ is the frequency shift of Raman scattering light from the combinatory frequencies $(\omega - \omega')$ and $(\omega + \omega')$ between the incident light frequency ω and that of the dipole oscillation ω'. The functions $J_0^{(2)}(\Delta\omega)$ and $J_1^{(2)}(\Delta\omega)$ are the SDFs of the normalized spherical harmonics $Y_0^{(2)}(\vartheta,\phi)$ and $Y_1^{(2)}(\vartheta,\phi)$. Accounting for the explicit form of the ACF given by Equation (3.50), we can write the SDF expression $J_m^{(2)}(\phi,\vartheta,\Delta\omega)$ in single crystals as [36, 64, 72]:

$$J_m^{(2)}(\phi,\vartheta,\Delta\omega) = \frac{5}{2\pi}\sum_{\alpha=1}^{3}\sum_{l=0}^{2} q_\alpha^{(2)} a_{\alpha lm}^{(2)}(\phi)\cos^{2l}\vartheta \,\frac{\omega_\alpha}{\omega_\alpha^2 + (\Delta\omega)^2} \tag{4.35}$$

where $m = 0$ and 1. The factors $a_{\alpha lm}^{(2)}(\phi)$ are given in Table 3.5 through Table 3.7. The symmetrized correlation frequencies of the HMM ω_α relate to the symmetrized correlation times τ_α by Equation (4.10).

In polycrystals, the SDF reduces to

$$J_m^{(2)}(\Delta\omega) = \frac{1}{2\pi}\sum_{\alpha=1}^{3} q_\alpha^{(2)} \frac{\omega_\alpha}{\omega_\alpha^2 + (\Delta\omega)^2} \tag{4.36}$$

In Figure 3.11 through Figure 3.15, the graphs of the initial values of the second-rank ACFs $K_m^{(2)} = K_m^{(2)}(\vartheta, q_1, t = 0)$ adapted to the cubic, hexagonal-trigonal, and tetragonal systems are shown. These graphs display also the angular dependence of the relative peak intensity of Rayleigh and Raman lines, polarized for $m = 0$ and depolarized for $m = 1$.

4.4 *Discussion and comparison with the experiment*

The microscopic effect of two symmetries in the intensities of the molecular spectral lines is investigated. The symmetry of the molecular motion affects the symmetrized correlation times τ_α(or frequencies ω_α). The site symmetry associates with the dynamical weights of the molecular motion hindered states $q_\alpha^{(v)}$. One can obtain the experimental values of τ_α and $q_\alpha^{(1)}$, for example, by studying the temperature dependence of the intensities of dielectric permittivity and/or infrared absorption as a function of the temperature for various single-crystal orientations. Under a HMM, the molecule dipole transforms itself with a certain probability, according to the irreducible representations of the crystallographic point symmetry group determined by the crystal structure. Therefore, we can assign a physical meaning to the weight of a hindered state $q_\alpha^{(1)}$: It determines the dipole-moment fraction that transforms with respect to the corresponding irreducible representation Γ_α. Similar outcomes can be taken from light scattering experiments. It should be remembered that the weights of the hindered states corresponding to perfect crystals are equal to the normalized dimensions of the appropriate irreducible representations.

A profound theoretical investigation of the effect of molecular thermal rotation in broadening infrared absorption lines was performed by Valiev in the framework of the SRD model [44]. It is helpful to compare the shape of the infrared absorption lines predicted in the frameworks of the two HMM models (SRD model and EAJ model). Both models show the same principal outcomes: A vibrational band represents a superposition of Lorentzian curves. The number of Lorentzians (1, 2, or 3) depends on the type of the molecular top in the SRD model or on the number of nonidentical, irreducible representations of the HMM symmetry group in the EAJ model.

The EAJ model presented by molecular rotational jumps to finite angles forming a group in the cubic system with single correlation time symmetrized on the three-dimensional irreducible representation conforms to the isotropic rotational diffusion of the spherical top molecules described by one constant of diffusion in the SRD model approach. A discrete HMM adapted to a group D_n ($n = 3, 4, 6$) with two correlation times symmetrized on two nonidentical, nonequivalent irreducible representations conforms to the rotational diffusion of symmetrical molecules with two diffusion constants.

Finally, a discrete HMM adapted to the group D_2 with three correlation times symmetrized on three nonidentical, irreducible representations relates to the asymmetrical molecule rotation with three diffusion constants [73].

We also observed the partial agreement of the normalization factors of the total weights of Lorentzians for both theoretical models. The weighting factors of Lorentzians are always normalized to unity in the SRD model. In the EAJ model, they are also normalized to unity if the HMM symmetry group does not include identical representations. Each point symmetry group of the axial rotation C_n (n = 2, 3, 4, 6) for the first-rank tensors includes an identical representation. The identical representations persist also in the second-rank tensors symmetrized on all point groups, except for the cubic system groups. Hence, in the frame of EAJ model, the normalizing number of the total relative intensity of some bands can be less than unity and the band intensities can show apparent (allowed by symmetry) lowering.

This effect can be detected by measuring the dielectric constant as a function of frequency in dielectric spectroscopy. If the HMM is adapted to a group that includes the identity representation, the theoretical signal of the reduced dielectric constant does not decrease to zero at high frequencies. In addition, the signal of the reduced loss factor does not attain its maximal value of 0.5 in this case. The graphs of Figure 4.7(f) shows this effect in $\eta'(\omega)$ and $\eta''(\omega)$ predicted for dielectric response from the molecules exhibited the HMM adapted to the axial groups C_n (n = 2, 3, 4, 6).

The significant divergence of the compared HMM models consists of the interpretation of the correlation times obtained from the half-widths of Lorentzian curves. In the theory using the SRD model, these correlation times depend on the rotational diffusion constants, whose basic idea has been forwarded from the hydrodynamic theory of viscosity. For large molecules or their associates, such an approach yields realistic outcomes. However, for small molecules rotating by means of jumps to finite angles, the SRD model approach did not show a satisfactory justification. Valiev has suggested a hypothesis that the correlation time of small molecules must be equal to the average time between two consecutive reorientations. In the framework of the EAJ model, this hypothesis has found full theoretical confirmation by the expression of the symmetrized correlation times, as shown in Equation (2.29). If every separate step of the motion has equal probability p_i, all symmetrized correlation times τ_α become equal to the average time between two successive steps of motion τ. In this case, an infrared absorption line consists of a unique Lorentzian and the half-width at half maximum (HWHM) of the spectral line is equal to the average frequency of the HMM.

In addition, the HWHM of a line for the small molecules measured by various spectroscopy techniques has been observed to be equal. The comparison of correlation times obtained from HWHM of infrared and Raman lines for valence vibrations of the asymmetrical molecules of water

HDO in the interval of temperatures 25°C to 85°C was performed [43]. At the initial times of observation, while the ACF decreases by 10 times for 85°C and by 20 times for 25°C, these correlation times were equal. Another proof of the equality of correlation times derives from the experimental half-widths of Rayleigh and infrared lines corresponding to the valence symmetrical vibrations of the triple bond C ≡ N in the molecules of acetonitrile CH_3CN at ambient temperatures up to 230°C [42]. These experimental data agree satisfactorily with the description of the HMM in the frame of EAJ model.

The anisotropic line intensities are predicted in single crystals. However, there is no anisotropy in the dielectric and infrared spectra for the HMM adapted to a group of the cubic system. In addition, no anisotropy of Rayleigh and Raman spectra is predicted for the HMM adapted to a group of the cubic system if the molecule occupies a regular cubic symmetry site.

The offered description of the molecular line shape is in accord with the experimental intensities of vibrational Raman modes in single crystals containing polyatomic ions. In Figure 4.10 and Figure 4.11, the

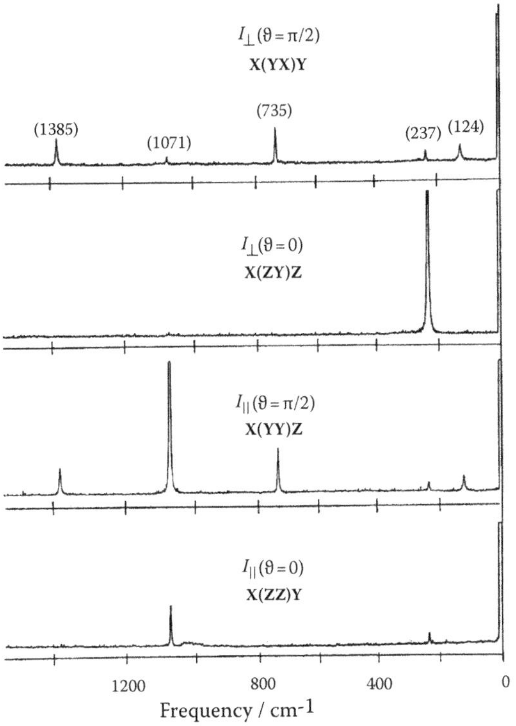

Figure 4.10 Raman spectra for various polarizations of incident light in a single-crystal of lithium nitrate, $LiNO_3$ [74].

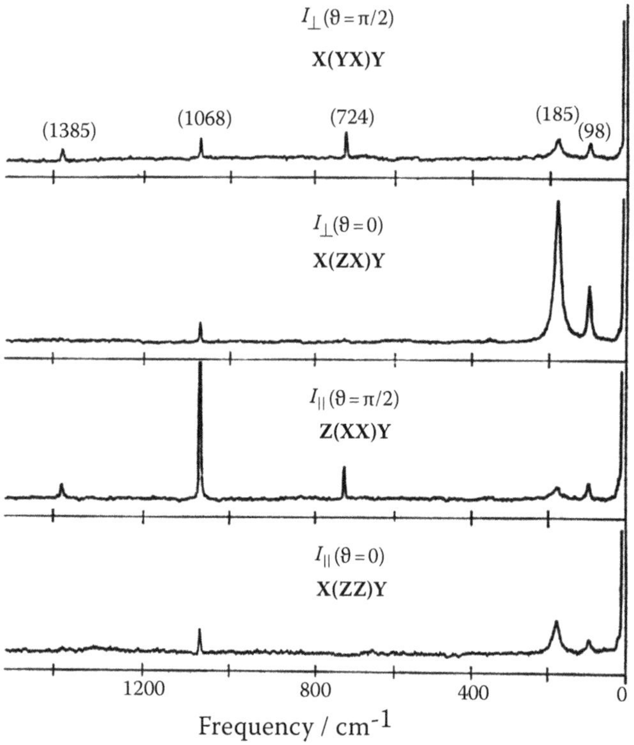

Figure 4.11 Raman spectra for various polarizations of incident light in a single-crystal of sodium nitrate $NaNO_3$ [75].

experimental polarized spectra of Raman scattering in single crystals of lithium nitrate ($LiNO_3$) and sodium nitrate ($NaNO_3$) for two orientations of these crystals are shown. These crystals have the structure of calcite with the space group D_{3d}^6 (Figure 4.12). The cations Na^+ and Li^+ occupy positions with the site symmetry S_6. The site symmetry of the NO_3^- anions is D_3. Three Raman-active vibrations are the nitrate-ion internal modes: $\nu_1(A_{1g})$, $\nu_3(E_g)$, and $\nu_4(E_g)$.

In the spectra, the symmetrical valence vibration of type A_{1g} of the anion dominates at the frequency 1071 cm^{-1} for $LiNO_3$ [74] and 1068 cm^{-1} for $NaNO_3$ [75] and gives rise to the intense polarized line. This line has maxima of intensity for perpendicular orientations of the crystals ($\vartheta = \pi/2$). Two other observable internal modes of the anion (at the left and the right of the first line) show weak lines of approximately equal intensity, whose polarized components $I_{||}$, as well as those depolarized $I_{\perp}$, have also maxima for perpendicular crystal orientations ($\vartheta = \pi/2$). Such angular behavior of scattered lines satisfies well to that of peak

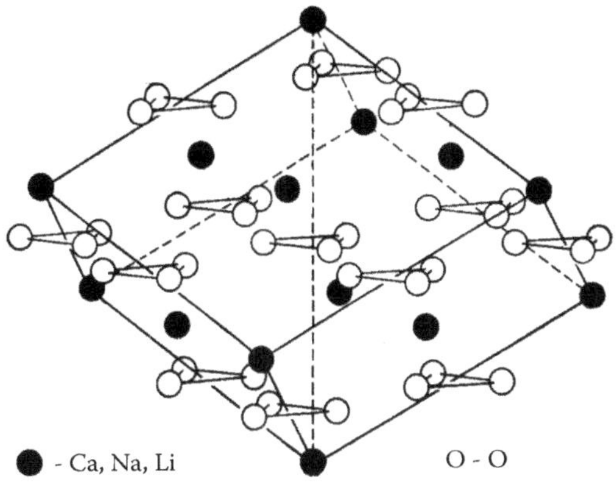

Figure 4.12 Calcite crystal structure [12].

values $J_0^{(2)}(\vartheta) \sim \sin^4 \vartheta$ and $J_1^{(2)}(\vartheta) \sim (1 - \cos^4 \vartheta)$ for the second-rank SDFs $J_0^{(2)}(\phi, \vartheta, \Delta\omega)$ and $J_1^{(2)}(\phi, \vartheta, \Delta\omega)$ expressed by Equation (4.35) and adapted to the groups of the hexagonal-trigonal crystallographic system taken with the values of dynamical weights $q_0 = 0$, $q_1 = 1$, and $q_2 = 0$. The respective graphs are shown in Figure 4.13.

In Figure 4.14, the graph of the Raman polarized line intensity is shown for the internal mode ν_1 (B_{1g}) = 1008 cm^{-1} in zircon $ZrSiO_4$ as a function of the angle ϑ between the crystal axis *a* and the direction of excitation *y*. The crystal is being turned around the *c* axis. The angular dependence of this line intensity follows the law $\cos^2 2\vartheta$. In Figure 4.15, the spectra are shown for four main polarizations [76]. The upper two spectra correspond to the polarized components of scattered lines with the optic axis **c** of the crystal oriented along the light vector **E** (**c** ⊥ **E**) and perpendicular to it (**c** ⊥ **E**).

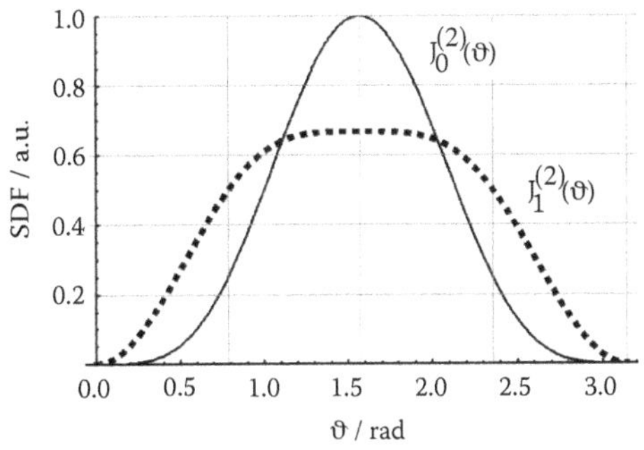

Figure 4.13 Angular dependence of the peak values $J_0^{(2)}(\vartheta)$ and $J_1^{(2)}(\vartheta)$, the second-rank SDFs adapted to a pure rotation point symmetry group of hexagonal-trigonal system. The fitted parameters are equal to $\Delta\omega = 0$, $q_0 = 0$, $q_1 = 1$, and $q_2 = 0$.

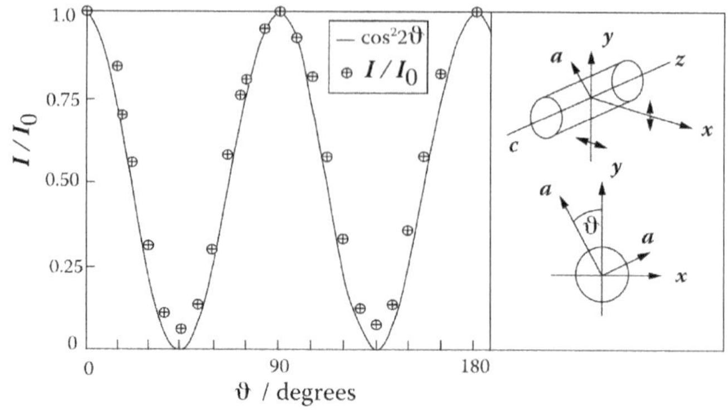

Figure 4.14 Intensity of the polarized line of Raman scattering of mode ν (B_{1g}) = 1008 cm^{-1} of zircon $ZrSiO_4$ presented as a function of the angle ϑ between the crystal *a* axis and the direction of excitation [76]. The sample is turned around the *c* axis. The predicted dependence of $\cos^2 2\vartheta$ in the tetragonal basis of the zircon crystal or of $\sin^2 2\vartheta$ in the cubic basis of the silicate-ion proves to be true.

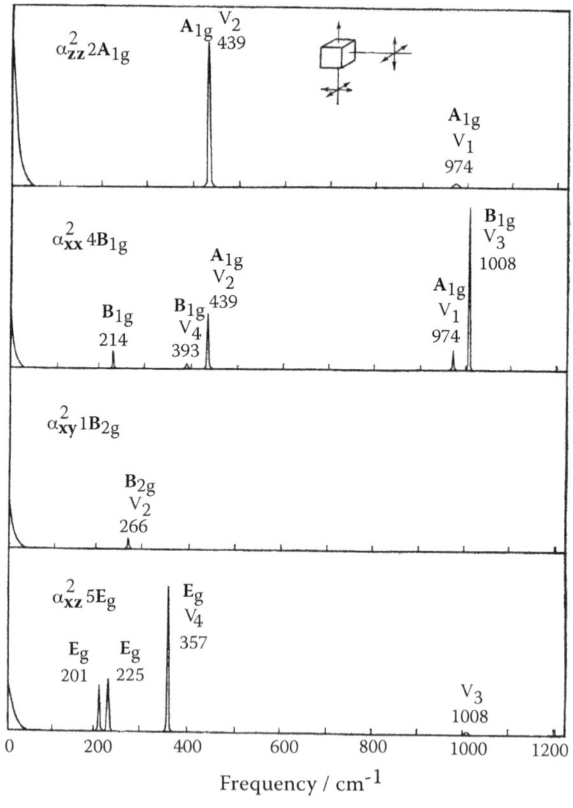

Figure 4.15 Raman spectra for various polarizations of incident light in the single-crystal of zircon $ZrSiO_4$ [76].

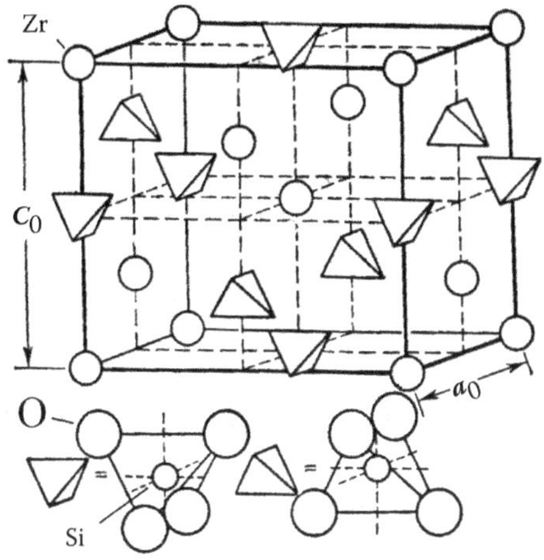

Figure 4.16 Zircon crystal structure [76].

The lower spectra are depolarized. The order of spectra is taken from top to bottom.

Zircon has a tetragonal unit cell containing two formula units $ZrSiO_4$ with the space symmetry group D_{4h}^{19} (Figure 4.16). Silicate-ion SiO_4^{4-} occupies places with the site symmetry D_{2d}. Our expressions for the spectral intensity relate to the reference frame, where the motion symmetry group is given. Hence, in order to describe the anisotropic properties of the polarized internal and rotational modes of vibration lines of the tetrahedral anions, we have to refer to the initial values of ACF $K_0^{(2)}(\vartheta, t = 0)$ adapted to a cubic group, whose reference frame is turned around its principal symmetry axis ($\vartheta = 0$) to the angle $\phi = \pi/4$. In the tetragonal position of the tetrahedral ion basis, it must be assumed that the dynamical weight of the state with the symmetry of the two-dimensional irreducible representation of cubic group is equal to zero, $q_1 \equiv q_1^{(2)} = 0$, and that symmetrized on three-dimensional one is equal to unity, $q_2 \equiv q_2^{(2)} = 1$. In that case, the dependence of ACF initial value $K_0^{(2)}(\vartheta, t = 0)$ of the polar angle ϑ is proportional to $\sin^2 2\vartheta$ (Figure 3.12). The graph shown in Figure 4.14 is drawn in the tetragonal basis of crystalline zircon. Transforming the ACF $K_0^{(2)}(\vartheta, t = 0)$ from the cubic basis to the basis taken in Figure 4.16 is carried out by replacing ϑ with $\vartheta + \pi/4$, which results in the proportionality of $\cos^2 2\vartheta$. This proves the correspondence of the presented theoretical discussion to the experimentally observed $\cos^2 2\vartheta$.

chapter five

Nuclear magnetic resonance spin-lattice relaxation applications

5.1 Generality of the nuclear magnetic resonance relaxation

One of the numerous experimental applications of nuclear magnetic resonance relaxation (NMR relaxation) is testing condensed media on the thermal motion of molecules [16–19]. Qualitatively, the effect produced by hindered molecular motion (HMM) to NMR relaxation consists of random time dependences of the energy transfer between nuclear Zeeman levels.

The theory of NMR relaxation describes the interacting processes inside the spin system, as well as between the spin system and other degrees of freedom of the substance. Within the limits of the validity of perturbation theory and the assumption that the correlation between more than two nuclear spins can be neglected, magnetic spin-lattice relaxation of identical nuclear spins $I = 1/2$ taking place due to the modulation of the intramolecular dipole–dipole interaction by the nuclear random motion is exponential.

This means that a single spin-lattice relaxation time characterizes the process of energy exchange between the macroscopic nuclear magnetization and the lattice. The relaxation time T_1 with respect to the static magnetic field is called the time of spin-lattice relaxation in the laboratory reference frame (LRF). The relaxation time $T_{1\rho}$ with respect to the oscillating magnetic field is called the relaxation time in the rotating reference frame (RRF). Magnetic relaxation of the quadruple nuclei with spins $I = 1$, arising due to modulation of the interaction between nuclear electrical quadruple moments and the oscillating gradient of the crystalline electric field by nuclear random motion, is also exponential [77].

For both kinds of relaxation, the expressions of relaxation rates T_1^{-1} and $T_{1\rho}^{-1}$ are usually presented by the linear combinations of the SDFs $J_m^{(2)}(\omega)$ of the second-rank unitary spherical tensor components $Y_m^{(2)}(g)$:

$$T_1^{-1} = A_I\left[J_1^{(2)}(\omega_0) + 4\,J_2^{(2)}(2\omega_0)\right] \tag{5.1}$$

and

$$T_{1\rho}^{-1} = A_I\left[\frac{3}{2}J_0^{(2)}(2\omega_1) + \frac{5}{2}J_1^{(2)}(\omega_0) + J_2^{(2)}(2\omega_0)\right] \tag{5.2}$$

The factor A_I, where $I = 1/2$ or 1, depends on the nature of relaxation events and equals respectively to

$$A_{1/2} = \frac{3\pi}{5}(N-1)\gamma^4\hbar^2 b^{-6} \tag{5.3}$$

or

$$A_1 = \frac{3\pi}{5}(e^2 qQ/\hbar)^2 \tag{5.4}$$

where the standard symbols are used. The resonance frequencies of nuclei in the LRF and the RRF are denoted respectively by $\omega_0 = \gamma B_0$ and $\omega_1 = \gamma B_1$ (B_0 is the intensity of the static magnetic field and B_1 is that of the rotating one). The factor γ is the nuclear gyro-magnetic ratio, $\hbar$ is the reduced Planck constant, b is the internuclear distance, N is the number of interacting nuclei, e is the elementary electronic charge, q is the modulus of the gradient of the axial-symmetry electric field, and Q is the constant of nuclear quadruple moment. The SDFs $J_m^{(2)}(\omega)$ relate to the autocorrelation functions $K_m^{(2)}(t)$ by means of Fourier transform:

$$J_m^{(2)}(\omega) = \int_{-\infty}^{\infty} K_m^{(2)}(t)\exp(i\omega t)\,\mathrm{d}t \tag{5.5}$$

With respect to Equation (3.50), the explicit form of the second-rank ACFs adapted to the extended angular jump (EAJ) model in single crystals, the required SDFs $J_m^{(2)}(\omega)$ are the functions of the variable quantities ϕ, ϑ, ω, τ_α, and $q_\alpha^{(2)}$ [36]:

$$J_m^{(2)}\left(\phi, \vartheta, \omega, \tau_\alpha, q_\alpha^{(2)}\right) = \frac{5}{2\pi}\sum_\alpha q_\alpha^{(2)} \frac{\tau_\alpha}{1+(\omega\tau_\alpha)^2}\left(\sum_{l=0}^{2} a_{\alpha l m}^{(2)}(\phi)\cos^{2l}\vartheta\right) \tag{5.6}$$

The variables ϕ and ϑ determine the geometry of the experiment, and the quantities ω, τ_α, and $q_\alpha^{(2)}$ determine its physics. The polycrystalline SDFs are the functions of the reduced number of variables ω, τ_α, and $q_\alpha^{(2)}$:

$$J_m^{(2)}\left(\omega,\tau_\alpha,q_\alpha^{(2)}\right)=\frac{1}{2\pi}\sum_\alpha q_\alpha^{(2)}\frac{\tau_\alpha}{1+(\omega\tau_\alpha)^2} \tag{5.7}$$

By substituting Equation (5.6) into Equations (5.1) and (5.2), we can obtain the general formulae T_1^{-1} and $T_{1\rho}^{-1}$ as functions of the NMR resonance frequencies ω_0 and ω_1, the angles ϑ and ϕ of the crystallographic reference frame (CRF) orientation in the LRF (static magnetic field), the dynamical weights $q_\alpha \equiv q_\alpha^{(2)}$, and the symmetrized correlation times τ_α. The explicit values of some factors and numbering data important in processing are given in Tables 3.1 through 3.3 and 3.5 through 3.8. In this chapter, we show how a new theoretical approach works by describing the NMR relaxation rates. As an example, we present the full discussion of the experimental data on the proton NMR relaxation in single-crystalline ammonium chloride. Then we improve the treatment of experimental data by studying the proton relaxation in symmetric three-atomic formations.

5.2 Relaxation through the motion adapted to the cubic symmetry groups

Because of multiplicity of the variables presented in monocrystalline expressions for relaxation rates, we prefer to derive only the reduced explicit formulas that satisfy the rationally chosen experimental conditions. In the framework of the EAJ model adapted to the symmetry group O or T, such a case is offered by the particular installation of a cubic unit cell with the polar angle ϑ taken equal initially to 0 with regard to the direction of the axial magnetic field intensity $\mathbf{B}_0(0,0,B_0)$. This allows us to install all principal symmetry axes C_2, C_3, and C_4 of the cube along the intensity $\mathbf{B}_0$ by turning the sample around the axis C_2, which is parallel to the direction [1,1,0] of the cubic unit cell ($\phi = \pi/4$).

At this condition, the relaxation rate T_1^{-1} is a function of six variables ϑ, ω_0, τ_1, τ_2, $q_1 = q_1^{(2)}$, and $q_2 = q_2^{(2)}$:

$$\begin{aligned}T_1^{-1} &= \frac{5}{24\pi}A_I\left\{3q_1\tau_1\left[\frac{1+2\cos^2\vartheta-3\cos^4\vartheta}{1+(\omega_0\tau_1)^2}+\frac{3-2\cos^2\vartheta+3\cos^4\vartheta}{1+(2\omega_0\tau_1)^2}\right]\right.\\ &\left.+2q_2\tau_2\left[\frac{1-2\cos^2\vartheta+3\cos^4\vartheta}{1+(\omega_0\tau_2)^2}+\frac{5+2\cos^2\vartheta-3\cos^4\vartheta}{1+(2\omega_0\tau_2)^2}\right]\right\}\end{aligned} \tag{5.8}$$

$T_{1\rho}^{-1}$ is the function of seven variables ϑ, ω_0, ω_1, τ_1, τ_2, $q_1 = q_1^{(2)}$, and $q_2 = q_2^{(2)}$:

$$T_{1\rho}^{-1} = \frac{5}{96\pi} A_I \left\{ 3q_1\tau_1 \left[\frac{3(1-3\cos^2\vartheta)^2}{1+(2\omega_1\tau_1)^2} + \frac{10(1+2\cos^2\vartheta - 3\cos^4\vartheta)}{1+(\omega_0\tau_1)^2} + \frac{3-2\cos^2\vartheta + 3\cos^4\vartheta)}{1+(2\omega_0\tau_1)^2} \right] + 2q_2\tau_2 \left[\frac{9(1+2\cos^2\vartheta - 3\cos^4\vartheta)}{1+(2\omega_1\tau_2)^2} + \frac{10(1-2\cos^2\vartheta + 3\cos^4\vartheta)}{1+(\omega_0\tau_2)^2} + \frac{5+2\cos^2\vartheta - 3\cos^4\vartheta}{1+(2\omega_0\tau_2)^2} \right] \right\} \tag{5.9}$$

The values of the factors $a_{\alpha lm}^{(2)}(\phi = \pi/4)$ are taken from Table 3.5. Expressions for relaxation rates T_1^{-1} and $T_{1\rho}^{-1}$ in polycrystalline samples derived by using Equations (5.1), (5.2), and (5.7) are

$$T_1^{-1} = \frac{1}{2\pi} A_I \left\{ q_1 \left[\frac{\tau_1}{1+(\omega_0\tau_1)^2} + \frac{4\tau_1}{1+(2\omega_0\tau_1)^2} \right] + q_2 \left[\frac{\tau_2}{1+(\omega_0\tau_2)^2} + \frac{4\tau_2}{1+(2\omega_0\tau_2)^2} \right] \right\} \tag{5.10}$$

and

$$T_{1\rho}^{-1} = \frac{1}{2\pi} A_I \left\{ q_1 \left[\frac{3}{2}\frac{\tau_1}{1+(\omega_1\tau_1)^2} + \frac{5}{2}\frac{\tau_1}{1+(\omega_0\tau_1)^2} + \frac{\tau_1}{1+(2\omega_0\tau_1)^2} \right] + q_2 \left[\frac{3}{2}\frac{\tau_2}{1+(\omega_1\tau_2)^2} + \frac{5}{2}\frac{\tau_2}{1+(\omega_0\tau_2)^2} + \frac{\tau_2}{1+(2\omega_0\tau_2)^2} \right] \right\} \tag{5.11}$$

Let us discuss Equations (5.8) through (5.11).

- Two regimes (fast and slow) correspond to the HMM with respect to the LRF. In the fast-motion regime, the conditions $\tau_\alpha \ll \omega_0^{-1}$, where $\alpha = 1$ and 2, are fulfilled, the angular variable ϑ vanishes in the formulae of relaxation rates given by Equations (5.8) and (5.9) and the anisotropy of T_1^{-1} and $T_{1\rho}^{-1}$ vanishes. Moreover, the relaxation rates in both the LRF and the RRF become equal to each other:

$$T_1^{-1} = T_{1\rho}^{-1} = \frac{5}{2\pi} A_I (q_1\tau_1 + q_2\tau_2) \tag{5.12}$$

- In the slow-motion regime ($\tau_\alpha >> \omega_0^{-1}$), both relaxations are anisotropic. However, if the HMM is adapted to an ideal cubic group and the probabilities of elementary steps of motion are equal each other [that is, $p(g_{ej}) = 1/\sigma$], both correlation times τ_1 and τ_2 become equal to the average time between two successive steps of motion τ ($\tau_1 = \tau_2 = \tau$) and neither anisotropy T_1^{-1} nor $T_{1\rho}^{-1}$ is predicted. Moreover, Equations (5.8) through (5.11) reduce to the famous formulae of Bloembergen-Purcell-Pound (BPP) derived in the frame of the model of isotropic rotational diffusion of linear and spherical molecules [17, 78]:

$$T_1^{-1} = \frac{1}{2\pi} A_I \tau \left\{ \left[1 + (\omega_0 \tau)^2\right]^{-1} + 4\left[1 + (2\omega_0 \tau)^2\right]^{-1} \right\} \tag{5.13}$$

and

$$T_{1\rho}^{-1} = \frac{1}{2\pi} A_I \tau \left\{ \frac{3}{2}\left[1 + (2\omega_1 \tau)^2\right]^{-1} + \frac{5}{2}\left[1 + (\omega_0 \tau)^2\right]^{-1} + \left[1 + (2\omega_0 \tau)^2\right]^{-1} \right\} \tag{5.14}$$

- All Equations (5.8) through (5.11) contain four dynamical variables q_1, q_2, τ_1, and τ_2, which are unknown in advance. They can be determined from single-crystalline experiments, if maxima exist in the curves of the temperature dependence of relaxation rates. Usual relaxation experiments are performed under conditions when the intensity of the static magnetic field is considerably higher than the intensity of the alternating magnetic field: $B_0 >> B_1$. In this case, the neighborhood of a maximum in $T_{1\rho}^{-1}$ falls to the region of the LRF slow-motion regime, and the strong inequality $\tau_\alpha >> \omega_0^{-1}$ is fulfilled. By neglecting the terms depending on ω_0, whose contribution to relaxation is small in this regime, Equation (5.9) for the relaxation rate $T_{1\rho}^{-1}$ reduces to

$$T_{1\rho}^{-1} = \frac{15}{32\pi} A_I \left[q_1 \tau_1 \frac{(1 - 3\cos^2\vartheta)^2}{1 + (2\omega_1 \tau_1)^2} + 2 q_2 \tau_2 \frac{1 + 2\cos^2\vartheta - 3\cos^4\vartheta}{1 + (2\omega_1 \tau_2)^2} \right] \tag{5.15}$$

- For some crystal orientations, Equation (5.15) simplifies extremely. This fact allows us to get unequivocal relationships between the dynamical weights and the minimal values of relaxation times in the RRF—in other words, to calculate the numbering values q_1 and q_2. For example:

$$\vartheta = 0^\circ, \quad \tau_1 = 0.5/\omega_1, \quad q_1 = 128\pi\omega_1/15 A_I T_{1\rho}{}^{(\mathrm{min})}(\vartheta = 0^\circ) \tag{5.16}$$

and

$$\vartheta = 54^\circ, \quad \tau_2 = 0.5/\omega_1, \quad q_2 = 32\pi\, \omega_1/5A_I\, T_{1\rho}^{(min)}\,(\vartheta = 54^\circ) \qquad (5.17)$$

5.3 Proton relaxation in crystalline ammonium chloride

5.3.1 Preamble

Ammonium chloride (NH_4Cl), one of the most studied substances, is used as a touchstone of the validity of various theories on the structure and physical properties of crystals. Its structure is assigned to a body-centered cube of CsCl type at ambient and lower temperatures [79]. Below 242.9 K, the crystal is in the ordered phase. The equilibrium disposition of its atoms is shown in Figure 5.1(a). The random local motion of ammonium ions does not change the ordered structure of the crystal. It means that the reorientation symmetry group of any ammonium ion vector is the point symmetry group of the tetrahedron T.

There are two equilibrium dispositions of the NH_4^+ tetrahedron with equal probability in the disordered phase observed at temperatures $T \geq$ 242.9 K [Figure 5.1(b)]. Hence the reorientation of any physical quantity follows the point symmetry group of the octahedron O in the disordered phase of NH_4Cl.

Experimental studies of the proton NMR relaxation times T_1 and $T_{1\rho}$ have been performed in ammonium chloride many times [27, 47–49, 80, 81]. Specific jump model [27], rotational diffusion (RD) model [80], and fixed angular jump (FAJ) model [47–49, 81] were satisfactorily applied to approximate the ammonium motion in powder samples.

In Figure 5.2, the temperature dependences of relaxation times T_1 and $T_{1\rho}$ along three main orientations of the single-crystal NH_4Cl are presented [27, 49]. Curves $T_1(T)$ and $T_{1\rho}(T)$ display the same temperature

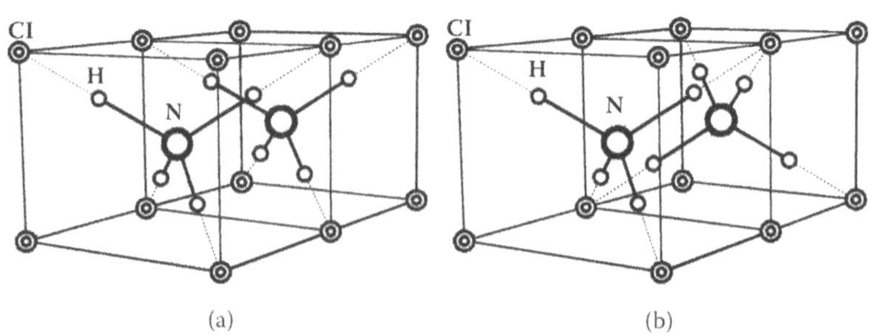

Figure 5.1 Crystalline structure of ammonium chloride NH_4Cl: (a) ordered phase; (b) disordered phase.

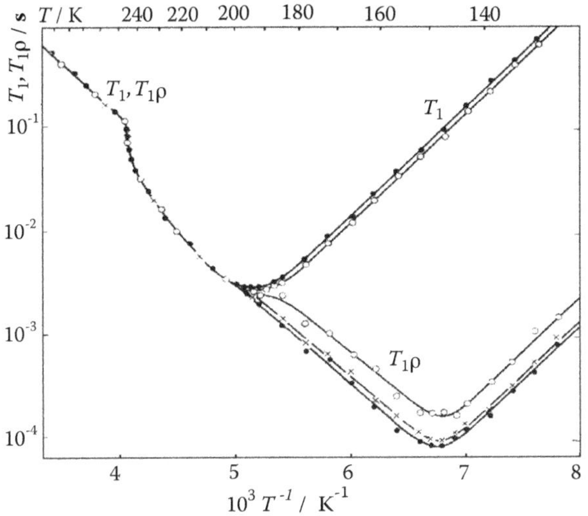

Figure 5.2 The proton spin-lattice relaxation times in the LRF (T_1) and the RRF ($T_{1\rho}$) drawn as a function of the inverse temperature in the single-crystalline ammonium chloride NH_4Cl for its three main orientations in the static magnetic field: o, $\mathbf{B}_0$ // [1,0,0]; ×, $\mathbf{B}_0$ // [1,1,0]; and •, $\mathbf{B}_0$ // [1,1,1]. The resonance frequency of protons in LRF $\nu_0 = \omega_0/2\pi = 14$ MHz, the intensity of rotating magnetic field $B_1 = 3.26$ mT [27].

behavior and do not show anisotropic properties at ambient temperature and below, down to the temperature region where T_1 achieves its minimal value. At the lowest temperatures, both relaxation times are anisotropic with the following properties: The relative anisotropy of $T_{1\rho}$ is more significant than that of T_1, and they have opposite anisotropy behavior.

Such anisotropic properties of curves cannot be derived in the frameworks of either the RD model or the FAJ model approaches. No anisotropy is predicted in the framework of the relaxation theory using the RD model. Using the FAJ model results in the anisotropy of T_1 and $T_{1\rho}$. However, it is not consistent with the experimental data. We present now the detailed discussion of these experimental data in the framework of relaxation theory using the EAJ model.

5.3.2 *Slow-motion regime of ammonium cations*

With respect to Equations (5.16) and (5.17), the experimental data on the minimal relaxation times measured for some orientations in the ordered phase of NH_4Cl allow us to calculate the dynamical weights q_1 and q_2 independently of each other. Experimental values $T_{1\rho}^{(min)}$ ($\vartheta = 54°$) = 89 mks and

$T_{1\rho}^{(min)}$ ($\vartheta = 0$) = 170 mks were obtained at the respective temperatures 148.5 K and 148 K in the field intensity $B_1 = 3.26$ mT [27]. The proton–proton distance b for NH_4^+ ions is taken by 1.695×10^{-10} m [79], the number of interacting protons is $N = 4$, the nuclear spin number is $I = 1/2$. The numerical values of dynamical weights calculated by using Equations (5.16) and (5.17) are equal to $q_1 = 0.25$ and $q_2 = 0.73$. Within the limits of experimental errors of 10% for $T_{1\rho}$, these data are satisfactorily consistent with the normalization condition given by Equation (2.4): $q_1 + q_2 = 0.98 \approx 1$.

The numerical values of the weights q_1 and q_2 for other temperature regions could be also determined if the corresponding minima of relaxation times is known. Fortunately, when the crystal structure remains unchanged, we do not have to recalculate the weights q_1 and q_2. Consequently, we take the values $q_1 = 0.25$ and $q_2 = 0.73$ as the constant values of these dynamical weights in the ordered phase of this crystal. In order to prove the validity of such an assumption, the results of calculations of $T_1^{(min)}$ and $T_{1\rho}^{(min)}$ by using Equation (5.15) and Equation (5.8) with the values $q_1 = 0.25$ and $q_2 = 0.73$ in single-crystalline NH_4Cl are presented in Table 5.1. The calculations performed for polycrystalline samples by using Equations (5.10) and (5.11) are included in the same table. Agreements to the experimental data are satisfactory.

In the LRF slow-motion regime, the condition $\tau_1 \approx \tau_2 \approx \tau >> \omega_0^{-1}$ is valid. By using this condition and replacing $q_2 = 1 - q_1$, Equations (5.8) and (5.15) reduce respectively to

$$T_1^{-1}(\vartheta, q_1) = \frac{5A_I}{32\pi\omega_0^2\tau}\left[2(3 - 2\cos^2\vartheta + 3\cos^4\vartheta) + q_1(1 + 10\cos^2\vartheta - 15\cos^4\vartheta)\right] \tag{5.18}$$

and

$$T_{1\rho}^{-1}(\vartheta, q_1) = \frac{15A_I\tau}{32\pi\left[1 + (2\omega_1\tau)^2\right]}\left[2(1 + 2\cos^2\vartheta - 3\cos^4\vartheta) - q_1(1 + 10\cos^2\vartheta - 15\cos^4\vartheta)\right] \tag{5.19}$$

The three-dimensional surface-plot graphs of theoretical relaxation rates $T_1^{-1}(\vartheta, q_1)$ and $T_{1\rho}^{-1}(\vartheta, q_1)$ are shown in Figure 5.3(a) and Figure 5.3(b). The graphs are drawn by using Equations (5.18) and (5.19) in arbitrary units as functions of the orientation ϑ and the dynamical weight q_1 of the two-dimensional irreducible representation E ($\alpha = 1$) for $\phi = \pi/4$. As regarding these graphs, we can see to what extent the theoretical anisotropy of the relaxation rates varies during the alteration of weight q_1 from 0 to 1. When the values of dynamical weights q_1 and q_2 become equal to

Table 5.1 Experimental and Theoretical Minimal Values of the Spin-Lattice Relaxation Times $T_1^{(min)}$ and $T_{1\rho}^{(min)}$ of Protons and Deuterons in Crystalline Ammonium Chloride NH_4Cl and ND_4Cl

		$T_1^{(min)}$ (ms)				$T_{1\rho}^{(min)}$ (mks)			
Nucleus	Crystal Orientation ϑ, degrees	ν_0 (MHz)	Experiment ±5%	Theory	Equation	B_1 (mT)	Experiment ±10%	Theory	Equation
Proton	0	14	2.78 [27]	2.76	(5.8)	3.26	170 [27]	$q_1 = 0.25$	(5.16)
	54	14	2.88 [27]	2.93	(5.8)	3.26	89 [27]	$q_2 = 0.73$	(5.17)
	90	14	2.85 [27]	2.88	(5.8)	3.26	100 [27]	101	(5.15)
	Powder	14	2.79 [49]	2.85	(5.10)	3.26	100 [49]	108	(5.14)
	Powder	14	—		—	1.61	53.6 [49]	53.3	(5.14)
	Powder	25	5.05 [80]	5.09	(5.10)	—	—	—	—
	Powder	18	3.51 [47]	3.66	(5.10)	1.19	39 [47]	39.4	(5.14)
Deuteron	Powder	9.5	0.46 [80]	0.44	(5.10)	—	—	—	—

Note: The angles $\vartheta = 0°$, $54°$, and $90°$ correspond to three main orientations of NH_4Cl single crystal [100], [111], and [110] in the static magnetic field $\mathbf{B}_0$. Experimental values $T_{1\rho}^{(min)}(\vartheta = 0°) = 170$ mks and $T_{1\rho}^{(min)}(\vartheta = 54°) = 89$ mks are used to calculate the dynamical weights $q_1 = 0.25$ and $q_2 = 0.73$. Reference numbers are given in brackets.

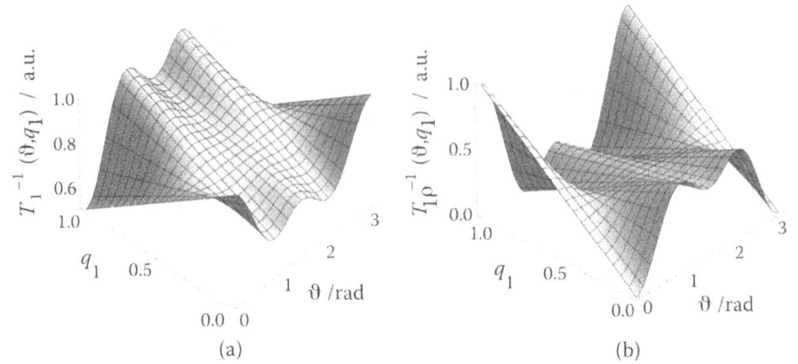

Figure 5.3 The theoretical (EAJ model) NMR spin-lattice relaxation rates drawn as functions of the polar angle ϑ and the dynamical weight q_1 for the cubic symmetry HMM: (a) $T_1^{-1}(\vartheta, q_1)$ and (b) $T_{1\rho}^{-1}(\vartheta, q_1)$. The azimuthal angle is $\phi = \pi/4$.

their static values $q_1 = 0.4$ and $q_2 = 0.6$, the anisotropy of relaxation rates vanishes. Straight lines, which are parallel to the axis ϑ, reflect the case of the absence of anisotropy for both $T_1^{-1}(\vartheta)$ and $T_{1\rho}^{-1}(\vartheta)$.

The two-dimensional graphs of the angular dependence $T_1^{-1}(\vartheta)$ and $T_{1\rho}^{-1}(\vartheta)$ for the three values q_1 are shown in Figure 5.4(a) and 5.4(b). The curves drawn with experimental value $q_1 = 0.25$ are shown by solid lines. The curves that correspond to $q_1 = 0$ and $q_1 = 1$ present the asymptotic angular dependences $T_1^{-1}(\vartheta)$ and $T_{1\rho}^{-1}(\vartheta)$ in the cubic sites.

5.3.3 *Fast-motion regime of ammonium cations*

In the temperature region where the condition for the fast-motion regime is fulfilled—that is, $\omega_0\tau_\alpha$, $\omega_1\tau_\alpha \ll 1$ ($\alpha = 1$ and 2)—the basic Equations (5.8) and

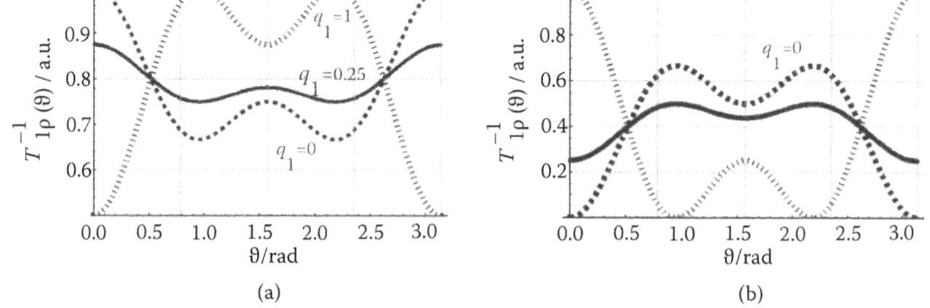

Figure 5.4 Theoretical intramolecular rates (EAJ model) of the nuclear magnetic spin-lattice relaxation drawn as a function of the main symmetry axis orientation ϑ for the cubic symmetry HMM: (a) $T_1^{-1}(\vartheta)$ and (b) $T_{1\rho}^{-1}(\vartheta)$. The fitted parameters are equal to $\phi = \pi/4$, $q_1 = 0.0, 0.25$, and 1.0. Note: $q_1 = 0.25$ is the value that satisfies the experimental curves shown in Figure 5.2.

(5.9) for both ordered and disordered phases of ammonium chloride reduce to Equation (5.12). Owing to the approximate equality of symmetrized correlation times ($\tau_1 \approx \tau_2 = \tau$), Equation (5.12) reduces to the simplest form

$$T_1^{-1} = T_{1\rho}^{-1} = \frac{5}{2\pi} A_I \tau \tag{5.20}$$

This theoretical prediction is consistent with the experimental data given in the temperature region above 200 K in Figure 5.2. Similarly, the experimental anisotropy of proton spin-lattice relaxation times has not been detected in the fast-motion regime of tetrahedral molecules of adamantane $C_{10}H_{16}$ in its monocrystalline plastic phase [56].

5.3.4 *Notes on the ammonium reorientation probabilities*

The prior discussion was developed by neglecting a small temperature shift of the minimal relaxation times relative to the various orientations of crystal NH_4Cl. Meanwhile, this experimental outcome can be used by studying ion microscopic dynamics.

Taking into consideration:

- the relationships between τ_α, τ, A_α, E_a, and p_i given by Equations (3.7) through (3.9),
- the explicit expressions A_1 and A_2 of the tetrahedron presented in Table 3.8,
- the normalizing condition for class probabilities $\Sigma\, p_i = 1$,
- the values $T_{1\rho}^{(min)}$ ($\vartheta = 54°$) = 89 mks and $T_{1\rho}^{(min)}$ ($\vartheta = 0$) = 170 mks [27],
- the condition of observing $T_{1\rho}^{(min)}$: $\omega_1\tau_1 = \omega_2\tau_2 = 0.5$,
- the average value of activation energy E_a = 4.83 kcal/mol (1 cal = 4.18 J) calculated from the slopes of the experimental curves, and
- the assumption that the probability of identical elementary rotation p(E) is the same as all others—that is, p(E) = $1/\sigma$, where $\sigma = 12$ is the order of the group T,

we can obtain the probabilities of the ammonium rotation in the different symmetry classes. They take the following numerical values in the ordered phase:

$$p(\mathrm{E}) = 0.083\ (0.083),\ p(\mathrm{C_3}) = 0.697\ (0.667),\ \text{and}\ p(\mathrm{C_2}) = 0.22\ (0.25)$$

The class probabilities derived under the condition of equal probability of any elementary rotations of the group T are presented in parentheses. It is also considered that the united class C_3 of the group T consists of eight elements. By using Equations (3.7) and (3.9), we can calculate the values of the symmetrized correlation times τ_α and the average one-fold probability transforms A_α in the ordered phase of NH_4Cl:

$$\tau_1 = 0.96\,\tau \quad \text{and} \quad \tau_2 = 1.01\,\tau$$
$$A_1 = -0.046 \quad \text{and} \quad A_2 = 0.01$$

These numerical data prove the validity of the assumption preliminarily adopted about the approximate equality of the symmetrized correlation times τ_1 and τ_2 to τ, the average time between two successive reorientations of the NH_4^+ ions.

5.3.5 *Motion and site symmetry of ammonium cations*

The magnetic dipole–dipole Hamiltonian is a symmetrical tensor of the second rank. With respect to the geometrical principles of invariance, the anisotropy of such a tensor is not prescribed in cubic crystals. According to Neumann's crystallographic principle, the apparent anisotropy of a physical quantity, being a tensor of the second rank, arises due to lowering (distorting) its cubic site symmetry [1–3]. The EAJ model allows us to investigate the symmetry distortion. With this goal, we use the new physical parameters introduced in the HMM theory: namely, a set of dynamical weights $q_\alpha^{(v)}$ of the hindered states symmetrized on the irreducible representations of the HMM group. In order to understand what kind of distortion takes place in a given case, we establish the phenomenological correspondence between the irreducible representations and the symmetry elements of the HMM group [82, 83].

The irreducible representation characters allocate to the symmetry elements of the group. The dynamical weight of an irreducible representation determines the probability that the molecular quantity transforms with the help of a full set of elements of the group by this irreducible representation. If the character of a class is equal to zero, there is no transform (change) of the molecular quantity under operators of the respective class symmetry elements in the given representation. Divergence of a dynamical weight from its static value corresponds to an increase or decrease of the contribution of such a class of elements into motion, which gives a nonzero character in the investigated representation. In other words, the increase of a class contribution to motion presents a test of changing the site symmetry toward the symmetry group for which this class is significant.

In the ordered phase of NH_4Cl, the dynamical weights were determined equal to $q_1 = 0.25$ and $q_2 = 0.73$, while static weights are equal, respectively, to 0.4 and 0.6. The characters of all classes in the two-dimensional representation E ($\alpha = 1$) of the tetrahedron group show nonzero values, but in the three-dimensional representation F ($\alpha = 2$), they have nonzero values only in the identical class 1E and in the class $3C_2$ (Table 3.1, Desk B). From this group-theoretical data, it follows that the increase of weight q_2 enlarges the contribution of the C_2 class (rotation by 180°) and, respectively, attenuates the contribution of the C_3 class (rotation by 120°) to the transformation of the second-rank tensors. The elements of class C_2 compose the basic elements of the point symmetry groups of the tetragonal crystallographic system (Table 3.1, Desk D). Class C_3 makes up a principal part of the point groups of the trigonal crystallographic system (Table 3.1, Desk H). Consequently, by taking into account the experimental values $q_1 = 0.25 < 0.4$ and $q_2 = 0.73 > 0.6$, we conclude that the tetragonal distortion of the tetrahedral site symmetry of ammoniums occurs in the ordered phase of NH_4Cl.

The fact of the absence of anisotropy of the relaxation times in the ordered phase for LRF fast-motion regime ($\omega_0\tau \ll 1$) is consistent with the dynamical principles of invariance in molecular motion. The high-symmetry fast motion averages both the local magnetic fields and the symmetry break effects to relaxation. In the disordered phase, the motion symmetry of tetrahedral ions NH_4^+ is octahedral, which is higher than that in the ordered phase of NH_4Cl. Therefore, no symmetry distortion manifests for NH_4Cl in its disordered phase.

Thus, the present study of the proton relaxation process in crystalline ammonium chloride confirms the known structural and dynamical data: The ammoniums occupy the cubic sites in the NH_4Cl unit cell (tetrahedral or octahedral), where they exhibit the hindered motion.

New structural knowledge consists of the microscopic properties of the hindered motion of ammoniums. Due to small energy thermal excitations, the instantaneous direction of an ammonium vector is not precisely known. One can determine only the limits of a solid angle where the ionic vector is directed with some probability density, which is periodic in crystals. With respect to the second-rank tensor molecular quantity, two hindered states relate to the ammonium cation in its orientational potential wells. One of them is two-fold degenerated state E and other is triply degenerated state F_2 or F. In the ordered phase, the motion symmetry group of rigid ammonium is tetrahedral with the values of dynamical weights $q_1 \equiv q(\mathrm{E}) = 0.25$ and $q_2 \equiv q(\mathrm{F}) = 0.73$. For comparison, the static weights are equal, respectively, to 0.4 and 0.6. Owing to the inequality $q(\mathrm{F}) > q(\mathrm{E})$, the tetrahedrons occupy the F state more probably than the E state. There are two reasons for such behavior of the ammonium cations. First, it occurs due to the inequality of the dimensions of the irreducible representations of the abstract point group of the tetrahedron—the geometrical nonequivalence. On the other hand, it happens owing to the symmetry distortion, which takes place in the ordered phase of NH_4Cl—the dynamical nonequivalence.

During thermoactivated motion, a vector of ammonium rotates to the angles, which are multiples of 120°, 180°, and 360° with respect to the point symmetry group of the tetrahedron in the ordered phase of NH_4Cl. The numerical data of class probabilities $p(\mathrm{E}) = 0.083$, $p(C_3) = 0.697$, and

$p(C_2) = 0.22$ allow us to note that C_3 class reorientation is more probable by comparison with C_2 class reorientation. This outcome is in accord with the ordered structure of the NH_4Cl crystal.

Dividing the class probabilities $p(E) = 0.083$, $p(C_3) = 0.697$, and $p(C_2) = 0.22$ to the corresponding class orders [$\sigma(E) = 1$, $\sigma(C_3) = 8$, and $\sigma(C_2) = 3$ (Table 3.1)], we can calculate the probabilities of the one-fold rotation. They are $p(E) = 0.083$, $p({}^1C_3) = 0.087$, and $p({}^1C_2) = 0.073$. We note that the probability of the unitary ${}^1C_3$ reorientation exceeds the probability of the unitary ${}^1C_2$ reorientation. This result is another justification to observing the ordering structure in the tetragonal distorted phase of NH_4Cl. Indeed, ${}^1C_2$ reorientation occurring in both phases of the substance is coaxial with the ${}^1C_4$ reorientation of the octahedral group. By virtue of the random character of the HMM, ${}^1C_2$ rotation can occasionally terminate by ${}^1C_4$ reorientation by giving rise to structural disorder. Therefore, a decrease of probability $p({}^1C_2)$ relative to its equiprobable value is consistent with the structure ordering. It should be noted that ${}^1C_3$ reorientation never breaks the considered ordered structure.

5.3.6 *Relaxation of protons and deuterons in powder ammonium chloride*

Experimental data of minimal values $T_1^{(min)}$ and $T_{1\rho}^{(min)}$ of protons and deuterons [47, 49, 80] are presented in Table 5.1. The results of their calculations by using Equations (5.10) and (5.11) with the help of the experimental data $q_1 = 0.25$ and $q_2 = 0.73$ are also placed in the same table. It is seen that the theoretical values $T_1^{(min)}$ and $T_{1\rho}^{(min)}$ are in agreement with the respective experimental values. In the limits of validity of the assumption about the equality of the probabilities of the elementary jumps, Equations (5.10) and (5.11) reduce, respectively, to Equations (5.13) and (5.14), which were derived earlier in the framework of the RD model approach; that is why the BPP theory is in good agreement with the experimental data taken in polycrystalline ammonium chloride as well as in other powder samples, where the highest symmetry hindered motion of atomic groups such as NH_4^+, BeF_4^{2-}, BF_4^{2-}, PF_6, $C_{10}H_{16}$, and so forth takes place [53, 54, 56, 78, 80, 84–91].

5.4 *Proton relaxation in three-atomic molecular fragments undergoing axial symmetry hindered motion*

Three-atomic formations—for instance, amine and methyl atomic groups NH_3 and CH_3—are attracting much attention from scientists, due to their notable contribution into the physical and chemical properties of inorganic and organic substances and proteins. The study of many substances with internal hindered motion of three-atomic fragments has been performed by using the proton magnetic resonance relaxation technique [31, 32, 91–113]. The theoretical discussion of the experimental data was carried out within the limits of the various HMM theories accounting for the RD as well as FAJ models. Meanwhile, conflicting discussions could not be considered satisfactory. We show in the following that some problems of quantitative description of the experimental data on proton relaxation therein could be resolved successfully in the framework of the EAJ model taken as the HMM model.

We are assuming that the motion symmetry group is the point symmetry group C_3 of the trigonal crystal system, the probabilities of the elementary rotations are equal to each other, and the symmetrized correlation time τ_α is equal to the average time τ between two successive orientational jumps.

Taking into account the explicit expressions of quantities $a^{(2)}_{\alpha lm}(\phi)$ and A_α presented in Tables 3.6 and 3.8, and by using Equations (5.1), (5.6), and (5.7), we obtain the expression describing the LRF relaxation rate T_1^{-1} adapted to the axial symmetry groups C_n ($n = 3, 4, 6$) in single-crystalline samples by:

$$T_1^{-1} = \frac{5}{8\pi} A_I \tau \left\{ q_1^{(2)} \left[\frac{1-\cos^4\vartheta}{1+(\omega_0\tau)^2} + \frac{1+6\cos^2\theta+\cos^4\vartheta)}{1+(2\omega_0\tau)^2} \right] \right.$$

$$\left. + q_1^{(2)} \left[\frac{1-3\cos^2\vartheta+4\cos^4\vartheta}{1+(\omega_0\tau)^2} + \frac{4(1-4\cos^4\vartheta}{1+(2\omega_0\tau)^2} \right] \right\} \qquad (5.21)$$

and in powdered samples by:

$$T_1^{-1} = \frac{1}{2\pi} A_I \tau \left(q_1^{(2)} + q_2^{(2)} \right) \left\{ \frac{1}{1+(\omega_0\tau)^2} + \frac{4}{1+(2\omega_0\tau)^2} \right\} \qquad (5.22)$$

where $q_1^{(2)}$ and $q_2^{(2)}$ are the dynamical weights of 2 two-dimensional equivalent, irreducible representations of the groups C_n ($n = 3, 4, 6$).

Experimental investigation of the NMR relaxation process performed in single-crystalline samples allows us to determine the numbering values of dynamical parameters. In accord with the experimental angular dependence of the proton relaxation rate T_1^{-1} [51, 112], the value of the weight $q_2^{(2)}$ must be taken equal to zero in Equation (5.21). This means that the hindered state with $\alpha = 2$ is useless in the HMM. Such a conclusion is consistent with the data taken from Raman spectra in single-crystalline $LiNO_3$ (Figure 4.10) and $NaNO_3$ (Figure 4.11).

By taking the value $q_2^{(2)} = 0$, Equations (5.21) and (5.22) reduce to

$$T_1^{-1} = \frac{5}{8\pi} A_I \tau q_1 \left[\frac{1-3\cos^2\vartheta+4\cos^4\vartheta}{1+(\omega_0\tau)^2} + \frac{4(1-\cos^4\vartheta)}{1+(2\omega_0\tau)^2} \right] \qquad (5.23)$$

and

$$T_1^{-1} = \frac{1}{2\pi} A_I \tau q_1 \left[\frac{1}{1+(\omega_0\tau)^2} + \frac{4}{1+(2\omega_0\tau)^2} \right] \tag{5.24}$$

where q_1 replaces $q_1^{(2)}$, for simplicity.

With $q_2^{(2)} = 0$ and taking into account the explicit expressions of quantities $a_{\alpha lm}^{(2)}(\phi)$ and A_α presented in Table 3.6 and Table 3.8, we obtain using Equations (5.2), (5.6), and (5.7) the expression describing the RRF relaxation rate $T_{1\rho}^{-1}$ adapted to the axial symmetry groups C_n (n = 3, 4, 6) in single-crystalline samples by:

$$T_{1\rho}^{-1} = \frac{5}{32\pi} A_I \tau q_1$$

$$\times \left[\frac{9(1-2\cos^2\vartheta+\cos^4\vartheta)}{1+(2\omega_1\tau)^2} + \frac{10(1-\cos^4\vartheta)}{1+(\omega_0\tau)^2} + \frac{1+6\cos^2\vartheta+\cos^4\vartheta}{1+(2\omega_0\tau)^2} \right] \tag{5.25}$$

and in polycrystalline samples:

$$T_{1\rho}^{-1} = \frac{1}{2\pi} A_I \tau q_1 \left[\frac{3/2}{1+(2\omega_1\tau)^2} + \frac{5/2}{1+(\omega_0\tau)^2} + \frac{1}{1+(2\omega_0\tau)^2} \right] \tag{5.26}$$

The principal novelty of Equations (5.23) through (5.26) consists of the additional factor q_1, the dynamical weight of the two-dimensional hindered state, which expands the power of the NMR relaxation technique by studying the HMM. The normalizing condition of the dynamical weights to unity allows us to determine the numbering limits of q_1: $0 \le q_1 \le 1$. In the case q_1 = 1, polycrystalline Equations (5.24) and (5.26) coincide with the antecedent expressions derived in the framework of BPP theory [17, 78]. However, this does not change the motion model. The case $q_1 = 1$ corresponds as well to EAJ model motion, but it is symmetrized only on the two-dimensional irreducible representation ($\alpha = 1$) of the group C_n ($n = 3, 4, 6$).

If the value q_1 satisfies the condition $q_1 < 1$, the motion is also jump-like with two corresponding hindered states; one of them is symmetrized on the two-dimensional irreducible representation (α = 1) and second is symmetrized on the identical representation ($\alpha = 0$). Because the HMM symmetrized on the identical representation does not transform a molecular physical quantity, it does not contribute to the relaxation. As a result, the relaxation rate decreases relative to its value predicted by the BPP approach.

The numerous experimental data prove the fact that the relaxation rate predicted by the BPP theory is faster than the experimentally observed

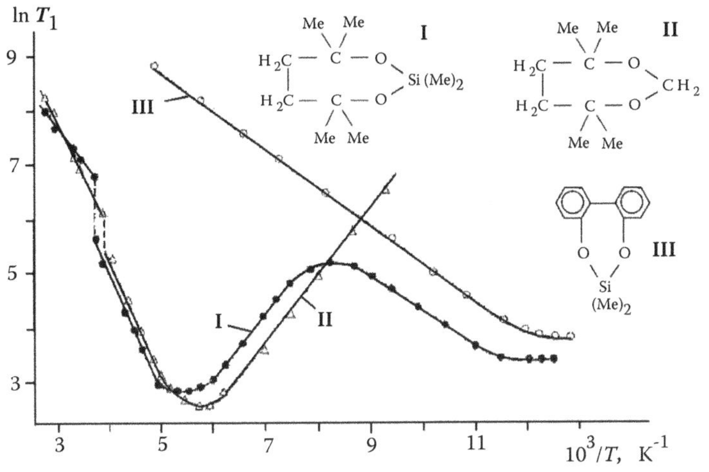

Figure 5.5 The temperature dependence ln $T_1 = f\ (10^3/T)$ of the proton spin-lattice relaxation times in three compounds I, II, and III [104] with rotating methyl groups: CH_3(I) $(CH_2)_2(CO)_2(CH_3)_4Si(CH_3)_2$; (II) $(CH_2)_2CH_2(CO)_2(CH_3)_4$; and (III) $(CH_2)_2Si(CH_3)_2(CO)_2(CH_3)_4$.

relaxation rate. Our experimental data on the times of proton relaxation in three compounds

I → 2,2,4,4,7,7-hexamethyl-2-sila-l,3-dioxepane $(CH_2)_2(CO)_2(CH_3)_4Si(CH_3)_2$
II → 4,4,7,7-tetramethyl-dioxepane $(CH_2)_2CH_2(CO)_2(CH_3)_4$
III → O,O′-diphendioxydimethylsilane $(CH_2)_2Si(CH_3)_2(CO)_2(CH_3)_4$)

where the methyl fragments exhibit the HMM, presented in Figure 5.5 [104]. We calculated the values of q_1 for given compounds by using Equation (5.24), which are smaller than 1 and are equal to $q_1(I) = 0.61$, $q_1(II) = 0.72$, and $q_1(III) = 0.62$. These data agree with our suggestion about the existence of two hindered states in the axial symmetry HMM symmetrized on the two-dimensional and identical irreducible representations in the axial symmetry HMM of a second rank tensor.

In Table 5.2, the data taken from the experiments performed by various authors in 46 substances of inorganic and organic compounds are collected. Experimental investigations have been performed in samples containing CH_3 molecular fragments. In Table 5.3, the data are collected on 17 solid compounds with the hindered motion of NH_3 groups. The database was created by using the available works and presents:

- Normalized values of experimental proton relaxation times taken in their temperature minimums $T_1^{(min)}$
- Times $T_1^{(min)}$ calculated according to the BPP theory

Table 5.2 Values of Experimental and Theoretical Times of the Proton Spin-Lattice Relaxation $T_1^{(min)}$, Dynamical Weights $q_1^{(2)}$, Resonance Frequencies ν_0, Proton–Proton Distances $b^{(HH)}$, and Activation Energies E_a of the Hindered Motion of CH_3 Molecular Fragments CH_3 in Powder Substances (1 kcal = 4.18 kJ)

No	Substances	Reference	$T_1^{(min)}$ reduced to one proton (ms)		$q_1^{(2)}$	Resonance frequency ν_0 (MHz)	H–H distance $b^{(HH)}$ (10^{-10} m)	Activation energy E_a (kcal/mole)
			Experiment	Theory BPP				
1	NH_3CHCH_3COO	[111]	12.86	11.3	0.87	27.5	1.78	—
2	ND_3CHCH_3COO	[111]	12.75	11.3	0.87	27.5	1.78	—
3	$NH_3CHCH(CH_3)_2COOH_2OHCl$	[111]	12.86	11.3	0.87	27.5	1.78	—
4	$NH_3CHCH_2CH_3COO$	[98]	13	11.3	0.87	27.5	1.78	—
5	$NH_3CH\ (CH_2)_2CH_3COO$	[98]	12.3	11.3	0.92	27.5	1.78	—
6	$NH_3CH\ (CH_2)_2CH_3COO$	[98]	14.4	11.3	0.78	27.5	1.78	—
7	$NH_3CH\ (CH_2)_3CH_3COO$	[98]	13.6	11.3	0.83	27.5	1.78	—
8	$C_{10}H_{21}NH_3Cl$	[109]	90	84.9	0.94	200	1.79	—
9	$Cl_3CC(CH_3)_2Cl$	[99]	18	10.74	0.60	25.3	1.79	2.5
10	$C(CH_3)_2(COOD)_2$	[108]	64	39.6	0.58	90	1.78	2.3
11	$(CH_3)_3CCN$	[91]	20.5	17.3	0.84	42	1.78	3.9
12	$(CH_3)_4NBr$	[93]	11	10.74	0.98	25.3	1.78	6.4
13	$(CH_3)_4NI$	[93]	13.3	10.74	0.81	25.3	1.78	5.3
14	$(CH_3)_4NCl$	[93]	12.5	10.74	0.86	25.3	1.78	6.8
15	$C_6C_6H_{18}$	[96]	11	10.74	0.98	25.3	1.78	1.9
16	C_2H_5OH	[102]	4.45	4.25	0.96	10	1.79	3
17	C_2H_5OD	[102]	4.8	4.25	0.89	10	1.79	3
18	CH_3PCl_2	[100]	25.5	13.2	0.52	30	1.80	2.25

19	CH_3PSCl_2	[100]	27	13.2	0.46	30	1.80	2.35
20	CH_3SCN	[100]	24	13.2	0.55	30	1.80	2.4
21	CH_3Cl	[105]	10.4	4.1	0.39	10	1.78	1.1
22	$(CH_3)_2CCl$	[93]	7.5	8.2	1.1	20	1.78	3.6
23	$(CH_2)_2(CO)_2(CH_3)_4Si(CH_3)_2$	[104]	9.5	5.75	0.61	14	1.78	1.3
24	$(CH_2)_2Si(CH_3)_2(CO)_2(CH_3)_4$	[104]	9.3	5.75	0.62	14	1.78	2.6–3.8
25	$(CH_2)_2CH_2(CO)_2(CH_3)_4$	[104]	8	5.75	0.72	14	1.78	2.6–3.8
26	$C_6H_4(N(CH_3)_2)_2$	[103]	17.1	12.7	0.74	29	1.797	—
27	$(CH_3)_3SiSi(CH_3)_3$	[92]	18.3	10.74	0.59	25.3	1.79	1.56
28	System: 1-CH_3, 9-CH_3	[106]	27	22	0.81	53	1.78	2.89
29	System: 1-CH_3, 9-CH_3	[106]	31	22	0.71	53	1.78	2.89
30	System: 1-CH_3, 9-CH_3	[106]	4.9	3.49	0.71	8.5	1.78	1.9
31	System: 1-CH_3, 9-CH_3	[106]	5.1	3.49	0.68	8.5	1.78	1.9
32	System: 1-DH_3, 9-CH_3	[106]	34	22	0.65	8.5	1.78	2.89
33	System: 1-DH_3, 9-CH_3	[106]	5.5	3.49	0.63	8.5	1.78	2.89
34	C_6H_{14}	[95]	25	20.5	0.82	50	1.78	1.1
35	C_7H_{16}	[95]	22.8	20.5	0.9	50	1.78	1.1
36	C_8H_{18}	[95]	22	20.5	0.93	50	1.78	1.1
37	$C_{10}H_{22}$	[95]	21	20.5	0.98	50	1.78	1.1
38	$C_{11}H_{24}$	[95]	21.8	20.5	0.94	50	1.78	1.1
39,40,41	$C_{12}H_{26}$, $C_{13}H_{28}$, $C_{14}H_{30}$	[95]	21.2	20.5	0.97	50	1.78	1.1
42, 43	$C_{16}H_{34}$, $C_{38}H_{78}$	[95]	22.9	20.5	0.9	50	1.78	1.1
44, 45	$C_{18}H_{38}$, $C_{28}H_{58}$	[95]	22.1	20.5	0.93	50	1.78	1.1
46	$C_{40}H_{82}$	[95]	21.9	20.5	0.94	50	1.78	1.1

Table 5.3 Values of the Experimental and Theoretical Times of the Proton Spin-Lattice Relaxation $T_1^{(min)}$, Dynamical Weights $q_1^{(2)}$, Resonance Frequencies ν_0, Proton–Proton Distances $b^{(HH)}$, and Activation Energies E_a of the Hindered Motion of NH_3 Molecular Fragments NH_3 in Powder Substances (1 kcal = 4.18 kJ)

No	Substances	Reference	$T_1^{(min)}$ reduced to one proton (ms) Experiment	$T_1^{(min)}$ reduced to one proton (ms) Theory BPP	$q_1^{(2)}$	Resonance frequency ν_0 (MHz)	H–H distance $b^{(HH)}$ (10^{-10} m)	Activation energy E_a (kcal/mole)
1	NH_3	[110]	14	12.2	0.87	42	1.68	2.41
2	$C_{10}H_{21}NH_3Cl$	[109]	68.8	58	0.84	200	1.68	—
3	$LiN_2H_5SO_4$	[113]	12	7.85	0.65	27	1.68	11.5
4	$N_2H_5H_2PO_4$	[107]	2.3	1.74	0.76	5.4	1.71	7.7
5	$N_2H_5H_2PO_4$	[107]	4.4	3.2	0.72	10	1.71	7.7
6	$N_2H_5H_2PO_4$	[107]	6.6	4.83	0.73	15	1.71	7.7
7	$(N_2H_5)_2HPO_4$	[107]	3.3	3.2	0.97	10	1.71	4.4
8	$(N_2H_5)_2HPO_4$	[107]	5.1	4.83	0.95	15	1.71	4.4
9	NH_3SO_3	[111]	12.5	8	0.64	27.5	1.68	—
10	NH_3CH_2COO	[111]	8.7	8	0.92	27.5	1.68	—
11	NH_3CHCH_3COO	[111]	9.9	8	0. 81	27.5	1.68	8.3
12	$NH_3CHCH(CH_3)_2COOH_2OHCl$	[111]	9.64	8	0.82	27.5	1.68	—
13	$NH_3CHHCOO$	[98]	8.4	8	0.95	27.5	1.68	6.5
14	$NH_3CHCH_2CH_3COO$	[98]	9.7	8	0.82	27.5	1.68	13
15	$NH_3(CH)_2(CH_3)_2COO$	[98]	10.35	8	0.77	27.5	1.68	6.5
16	$NH_3CH\ (CH_2)_2CH_3COO$	[98]	10.9	8	0.73	27.5	1.68	7.6
17	$NH_3CH\ (CH_2)_3CH_3COO$	[98]	9.2	8	0.87	27.5	1.68	9.2

- Values of dynamical weights q_1 calculated by using Equation (5.24) with the help of the experimental times $T_1^{(\min)}$
- Frequency of the proton resonance $\nu_0 = \omega_0/2\pi$
- Proton–proton distance b^{HH}
- HMM activation energy E_a

The data $T_1^{(\min)}$ of Table 5.2 present the values normalized to one proton in the appropriate molecule. Comparing the values of dynamical weight q_1, we can note that these values are enclosed mainly in the interval of numbers $0.6 \le q_1 \le 0.98$; and q_1 exceeds 1.0 only in the case of the Sample 22 of Table 5.2. Hence in the majority of samples, the condition $q_1 < 1$ is fulfilled. This means that the HMM symmetrized on the two-dimensional representation dominates in the HMM of methyl and amine fragments. At the same time, there is also the HMM symmetrized on the identical representation. The least values of q_1 ($q_1 < 0.8$) are observed in several substances where the relaxation process is not usually exponential and therefore these cases are to be discussed elsewhere.

As to the concluding remarks about the HMM of methyl and amine atomic systems, we can note:

- The hindered motion of such atomic systems takes place by means of angular jumps of molecular vectors to the angles, which are multiples of 120°.
- The apparent increase of relaxation times (decrease of relaxation rates) comparatively to those given by the BBP theory arises because the hindered states of the fragments are symmetrized on both irreducible representation of the group C_3: first, the two-dimensional one, which is effective in relaxation; and second, the identity representation, which is useless in relaxation.

chapter six

Incoherent neutron scattering applications

6.1 Preamble

In this preamble, we present introductory knowledge on the incoherent neutron scattering phenomenon in order to show the role assigned to the hindered molecular motion (HMM) therein. Our intercession is based on publications describing the theory and the instrumental technique of neutron scattering by matter (for example, [20–24, 114, 115]).

Neutron scattering is an experimental technique intended for the investigation of the structure and dynamics of condensed matter. This knowledge is accessible by analyzing energy and momentum transfer between scattered neutrons and the probed sample. A neutron is a microparticle with spin $s = 1/2$ and magnetic moment $\mu = -1.923\mu_N$, where $\mu_N = e\hbar/2m_p$ is a nuclear magneton. It has no charge, and its electric dipole moment is either zero or too small to be measured. For these reasons, neutrons can penetrate matter far better than charged particles (electrons, protons, ions, and so on). Furthermore, neutrons interact with atoms via nuclear rather than electrical forces, and nuclear forces are very short range—on the order of a few Fermis (1 Fermi = 10^{-15} m). Thus, as far as the neutron is concerned, solid matter is not very dense because the size of a scattering center (nucleus) is typically 100,000 times smaller than the distance between such centers. Therefore, neutrons can travel large distances through most materials without being scattered or absorbed.

A free neutron decays into a proton, an electron, and an antineutrino with a lifetime of about 15 minutes. Neutrons can be described either as classical particles with a mass of $m = 1.675 \times 10^{-27}$ kg or via the de Broglie formalism as a wave whose wavelength λ relates to the wave-vector **k** and the neutron velocity **v** as:

$$\lambda = \frac{2\pi}{|\mathbf{k}|} = \frac{h}{m|\mathbf{v}|} = \frac{h}{mv} \tag{6.1}$$

Table 6.1 Classification of Neutrons by Their Range of Energy, Corresponding Temperature, and Wavelength

	Hot	Thermal	Cold
Temperature T (K)	1000–6000	60–1000	1–120
Energy E (meV)	100–500	5–100	0.1–10
Wavelength λ (Å)	1–0.4	4–1	30–3

Note: 1 meV = 3.2×10^{-16} J, 1 Å = 10^{-8} m

with the Planck constant $h = 2\pi\hbar = 6.626 \times 10^{-34}$ Js. The energy E of a free neutron then is

$$E = \frac{1}{2}mv^2 = \hbar\omega = h\nu = \frac{\hbar^2 k^2}{2m} = \frac{h^2}{2m\lambda^2} \tag{6.2}$$

Neutrons are classified by their energy (hot, thermal, or cold), and their average energy corresponds to about $k_B T$ with average temperatures $T \approx 2000$ K, 300 K, or 25 K, respectively. The range of neutron energies E and wavelengths λ corresponding to this rough classification are presented in Table 6.1.

At the same time, the neutron energy E is in the energy range of excitations in condensed matter and the wavelength λ is comparable to interatomic distances. This fortunate match, for both timescale and spacing, makes neutrons an excellent exceptional probe for condensed matter structure and dynamics.

Owing to the uncharged nature of the neutron—contrary to an electromagnetic radiation and the electron—the neutron does not interact with the atomic charge distribution but rather directly with the atomic nucleus (via the strong nuclear force) and with magnetization density fluctuations (via their magnetic moment $\boldsymbol{\mu}$). The case of magnetic scattering will not be subject of further discussion here as it is beyond the scope of this investigation.

The range of the nuclear force action is orders of magnitude smaller than the neutron wavelength, and therefore the scattered wave is isotropic with its amplitude proportional to the so-called scattering length b. The value of b does not only depend on the element but also on its isotope and the spin state of the neutron–nucleus system. In the absence of a theory for nuclear forces, b is an experimentally determined quantity. Contrary to x-rays where the scattering strength increases with the number of electrons, b varies in a random manner from element to element. For the thermal and cold neutrons, b is practically independent of the incident neutron energy.

A general neutron scattering experiment consists of measuring the energy transfer ΔE and the momentum transfer $\Delta \mathbf{k} = \mathbf{Q}$ between sample and neutron, visualized in Figure 6.1.

Incoming neutrons with an incident wave vector $\mathbf{k}_0$, wavelength λ_0, and energy E_0 interact with the sample, recoil off and are detected at an

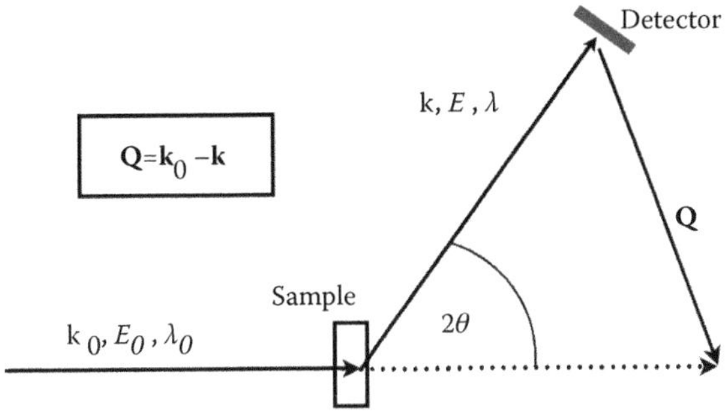

Figure 6.1 Schema of a scattering experiment.

angle 2θ with a final **k**, λ, and *E*. The scattering vector **Q** is defined as the change in wave vector: $\mathbf{Q} = \mathbf{k}_0 - \mathbf{k}$. By the way,

$$\Delta E = E_0 - E = \hbar\omega = \frac{\hbar^2}{2m}\left(k_0^2 - k^2\right) \tag{6.3}$$

and

$$\mathbf{Q} = \mathbf{k}_0 - \mathbf{k} = \Delta\mathbf{k} \tag{6.4}$$

These two quantities, ΔE and $\Delta\mathbf{k} = \mathbf{Q}$, include all the information needed about the structure and dynamics of the sample.

When neutrons are scattered by substance, the process can alter both the momentum **k** and the energy *E* of the neutrons and the matter. Scattering with no change in the energy of the incident neutron—or, in terms of the wave vector of the neutron, scattering in which the direction of the vector **k** changes but not its magnitude—is called elastic scattering [Figure 6.2(a)]. Alternatively, scattering in which the exchange of energy and momentum between the incident neutron and the sample causes both the direction and the magnitude of the neutron's wave vector to change is called inelastic scattering [Figure 6.2(b) and Figure 6.2(c)].

A scattering by matter is not necessarily elastic, as it is for a single, rigidly fixed nucleus, because atoms in matter are free to move to some

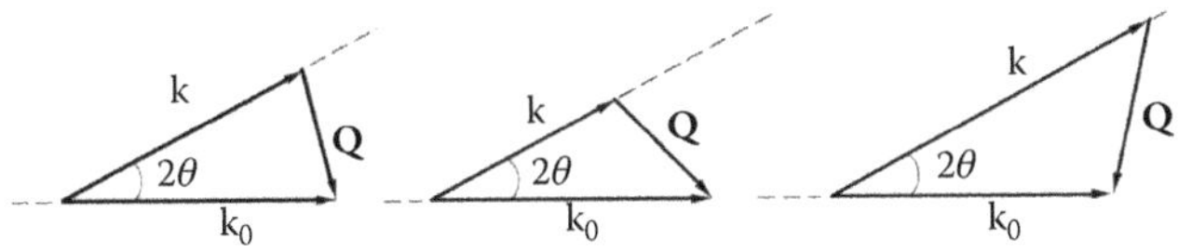

Figure 6.2 Scattering triangles: (a) for an elastic scattering event with $\mathbf{k} = \mathbf{k}_0$, (b) for an inelastic scattering events with $\mathbf{k} < \mathbf{k}_0$, (c) for an inelastic scattering event with $\mathbf{k} > \mathbf{k}_0$.

extent. They can therefore recoil during a collision with a neutron, or if they are moving when the neutron arrives, they can impart or absorb energy. As is usual in a collision, the total momentum and energy are conserved: When a neutron is scattered by matter, the energy ΔE lost by the neutron is gained by the sample. The vector relationship ($\mathbf{Q} = \mathbf{k}_0 - \mathbf{k}$) between $\mathbf{Q}$, $\mathbf{k}_0$, and $\mathbf{k}$ displayed pictorially by Figure 6.2 is called a scattering triangle. This triangle also emphasizes that the magnitude and direction of the scattering vector $\mathbf{Q}$ are determined by the magnitudes of the wave vectors for the incident and scattered neutrons and the angle 2θ through which a neutron is deflected during the scattering process.

Generally, 2θ is referred to as the scattering angle. It follows for elastic scattering [Figure 6.2(a)] that $k_0 - k = 0$ and $\Delta E = 0$. A little trigonometry applied to the scattering triangle shows us $Q = 2k \sin\theta$ or, by virtue of $k = k_0 = 2\pi/\lambda$, $Q = 4\pi \sin\theta/\lambda$. This univocal relation between Q and θ simplifies using a neutron-scattering spectroscopy technique: A study of a physical quantity dependence of wave-vector transfer Q (in the limits of $0 - 2k$) is performed by altering the scattering angle θ.

In all neutron-scattering experiments, scientists measure the intensity of neutrons scattered by matter (per incident neutron) as a function of the variables $\mathbf{Q}$ and $\Delta E = \Delta(\hbar\omega)$. This scattered intensity, denoted $S(\mathbf{Q}, \omega)$, is often referred to colloquially as the neutron-scattering law or the dynamic structure factor for the sample. In a complete and elegant analysis, Van Hove showed that the scattering law could be written exactly in terms of time-dependent correlations between the positions of pairs of atoms in the sample [114]. This result implies that $S(\mathbf{Q}, \omega)$ is simply proportional to the Fourier transform of a function that gives the probability of finding two atoms a certain distance apart. It is the simplicity of this result that is responsible for the power of neutron scattering. If nature had been unkind and included correlations between triplets or quadruplets of atoms in the expression for the scattering law, neutron scattering could not have been used to directly probe the structure of materials.

The type of scattering in which an incident neutron wave (all neutrons of the incident beam) interacts with all the nuclei in a sample in a coordinated fashion—that is, the scattered waves from all the nuclei have definite relative phases and can thus interfere with each other—is called coherent scattering. Incoherent scattering is the scattering in which an incident neutron wave interacts independently with each nucleus in the sample—that is, the scattered waves from different nuclei have random, or indeterminate, relative phases and thus cannot interfere with each other.

Van Hove's result provides a way of referring the intensity of the scattered neutrons to the relative positions and the relative motions of atoms in matter. In fact, his formalism is a tool to reveal the scattering effects of two types. The first is coherent scattering in which the neutron wave interacts with the whole sample as a unit so that the scattered waves from different nuclei interfere with each other. This type of scattering depends on the relative dispositions of the constituent atoms and thus gives information about the structure of materials. The simplest type of coherent neutron scattering is neutron diffraction or Bragg scattering. Elastic coherent scattering (with $\Delta E = 0$) allows us to learn about the equilibrium structure, whereas inelastic coherent scattering (with $\Delta E \neq 0$) provides information about the collective motions of the atoms, such as those that produce vibrational waves in a crystalline lattice. In the second type of scattering, incoherent scattering, the neutron wave interacts independently with each nucleus in the sample so that the scattered waves from different nuclei do not interfere. Rather the intensities from each nucleus just add up. For example, incoherent scattering may be due to the interaction of a neutron wave with the same atom but at different positions and different times, thus providing dynamic and static information about atomic random motion during the thermal-activated translational and rotational movement of molecules, chemical exchange, and conformational transition.

The incoherent scattering function $S(\mathbf{Q}, \omega)$ depends on the radius vector $\mathbf{r}(t)$ of the nucleus scatterer, the position vector, which can be separated into two different components, reflecting the diverse kinds of motions

$$\mathbf{r}(t) = \mathbf{r}_{eq}(t) + \mathbf{u}(t) \tag{6.5}$$

where $\mathbf{u}(t)$ is the displacement vector of the nucleus from its equilibrium position inside the molecule due to the internal vibrations (zero-point motion) and $\mathbf{r}_{eq}(t)$ describes the instantaneous location of this equilibrium position at a time t, with respect to a laboratory reference frame (LRF). Thus $\mathbf{r}_{eq}(t)$ can describe the random motion of whole molecule and, in its time, it can be separated into two parts

$$\mathbf{r}_{eq}(t) = \mathbf{r}_{eq}^{trans}(t) + \mathbf{r}_{eq}^{rot}(t) \tag{6.6}$$

where $\mathbf{r}_{eq}^{trans}(t)$ is related to translational random motion of the whole molecule and $\mathbf{r}_{eq}^{rot}$ is related to rotational random motion of the molecular vector around its point of application.

In liquids, translational diffusion of the molecule can occur. That is, the term describing translational random motion is time dependent. However, the focus of this work is on the properties of local HMM.

In the case of a solid substance, the center of mass of the molecule is restricted to well-defined positions $\mathbf{r}_0^{trans}$, and thus translational displacement usually does not occur on a timescale relevant for experimental observations. Around these well-defined positions, small fluctuations mapped here by $\mathbf{u}_{fluct}^{trans}(t)$ occur nevertheless as a result of thermal agitation. Therefore, for solid sample vector $\mathbf{r}_{eq}^{trans}(t)$, one can write

$$\mathbf{r}_{eq}^{trans}(t) = \mathbf{r}_0^{trans} + \mathbf{u}_{fluct}^{trans}(t) \tag{6.7}$$

When the molecule rotates around its center of mass, it is very convenient to describe the rotation of internuclear vectors $\mathbf{r}_0^{rot}(t)$ by instantaneous jumps between several angular potential wells. Moreover, during a long-term interval between the jumps, the molecule occupies an orientation where small-amplitude rotational oscillations (librations) occur that depend on intermolecular interactions. Therefore,

$$\mathbf{r}_{eq}^{rot}(t) = \mathbf{r}_0^{rot}(t) + \mathbf{u}_{libr}^{rot}(t) \tag{6.8}$$

$\mathbf{r}_0^{rot}(t)$ is time dependent if several potential wells exist and time independent if only one precise equilibrium orientation occurs. $\mathbf{u}_{libr}^{rot}(t)$ is the deviation vector corresponding to the librations.

Finally, for a solid sample, the position vector of a scattered neutron $\mathbf{r}(t)$ can be expressed as

$$\mathbf{r}(t) = \mathbf{u}_{fluct}^{trans}(t) + \mathbf{r}_0^{rot}(t) + \mathbf{u}_{libr}^{rot}(t) \tag{6.9}$$

In summary:

- Neutrons interact directly with atomic nuclei so that a neutron scattering spectrum includes full knowledge of the substance structure and its internal motion.
- Incoherent neutron scattering is scattering in which an incident neutron wave interacts independently with each nucleus in the sample; that is, the scattered waves from different nuclei have random, or indeterminate, relative phases and thus cannot interfere with each other. This is a case for application of the HMM model.

The main problem of the neutron scattering method of scientific research deals with the assignment of various contributions to experimental data. For this purpose, molecular crystals present a convenient species where the translational motion of atoms is considerably frozen and the local hindered motion of nuclear vectors gives rise to the main contribution to incoherent scattering. At the same time, significant difficulty of the spectral line-shape description in molecular crystals remains. In this chapter, the extended angular jump (EAJ) model is used to describe the spectra of incoherent neutron scattering in molecular crystals. Practical applications of experimental incoherent scattering intensities in several powder samples are given, comparing the function of different HMM models.

6.2 Basis of the theory of incoherent neutron scattering

The starting point of the theory of incoherent neutron scattering is the expression of the double differential cross-section $d^2\sigma(\mathbf{k},\mathbf{k}_0)$ of the neutron beam scattered from the substance (target) in unit time [20–24, 114, 115]. It is equal to the number of neutrons enclosed in an element of solid angle $d\Omega$ directed along the momentum vector (wave vector) **k** of scattered neutrons, in which the energy is enclosed in the interval of values between $\hbar\omega$ and $\hbar(\omega + d\omega)$:

$$d^2\sigma(\mathbf{k}_0,\mathbf{k}) = \frac{k}{k_0}\left[\sigma_{coh}S_{coh}(\mathbf{Q},\omega) + \sigma_{inc}S_{inc}(\mathbf{Q},\omega)\right]d\Omega d\omega \tag{6.10}$$

where $\mathbf{k}_0$ is the momentum vector of the incident neutrons, $\mathbf{Q} = \mathbf{k}_0 - \mathbf{k}$ is the momentum transfer vector (wave-vector transfer), S_{coh} and S_{inc} are the coherent and incoherent scattering functions, and σ_{coh} and σ_{inc} are the respective scattering cross-sections. Scattering cross-sections σ_{coh} and σ_{inc} relate to scattering length b by a different manner, namely,

$$\begin{aligned} \sigma_{coh} &= 4\pi(\bar{b})^2 \\ \sigma_{inc} &= 4\pi[(\bar{b}^2) - (\bar{b})^2] \end{aligned} \tag{6.11}$$

The first part of Equation (6.11) provides the correlations between the positions of various nuclei at different times, and the second part only gives the correlation between the positions of the same nuclei at different times. Contrary to incoherent scattering, the waves scattered from various atoms of the same crystal can interfere in coherent scattering. This can be explained by the fact that the neutron interacting with the sample does not see a crystal of uniform scattering potential. Instead, the scattering varies from one point to the next one in the sample. Only the average scattering potential can lead to interference effects; therefore, $(\bar{b})^2$ can be used to describe coherent scattering. The deviations from the average potential are randomly distributed, and thus they cannot lead to interference effects; therefore, incoherent scattering is proportional to the mean square deviation.

The different scattering lengths of various atoms and isotopes result in different scattering cross-sections. The magnitude of cross sections for

Table 6.2 Thermal-Neutron Scattering Cross Sections of Some Elements [115]

Element	Symbol	σ_{coh} / barns	σ_{inc} / barns
Hydrogen	H	1.7599	79.91
Deuterium	D	5.597	2.04
Carbon	C	5.554	0.001
Nitrogen	N	11.01	0.49
Boron	B	3.54	1.70

Note: 1 barn = 10^{-28} m^2

several nuclei is displayed in Table 6.2. We can see that hydrogen ^{1}H has a considerably larger incoherent cross section compared to all the other elements. That is why hydrogenous materials are suitable for study by incoherent neutron scattering techniques.

The coherent neutron scattering function S_{coh} is the subject of interest in the neutron diffraction physics dealing with static events in scattering phenomena. The random motion of nuclei contributes only to the incoherent neutron scattering function (INSF) S_{inc}. In conformity with the aim of this work, the INSF attracts our further particular attention. INSF is the Fourier transform of the intermediate scattering function (ISF) I_{inc}

$$S_{inc}(\mathbf{Q},\omega)=\frac{1}{2\pi}\int_{-\infty}^{\infty} I_{inc}(\mathbf{Q},t)\exp(-\mathrm{i}\omega t)\,\mathrm{d}t \tag{6.12}$$

In its turn, the ISF presents the autocorrelation function (ACF) of the normalized neutron wave function $\mathrm{F}(\mathbf{Q},\mathbf{b})=\exp[i\mathbf{Q}\mathbf{b}(t)]$:

$$I_{inc}(\mathbf{Q},t)=\langle\exp[-\mathrm{i}\mathbf{Q}\mathbf{b}(t)]\exp[\mathrm{i}\mathbf{Q}\mathbf{b}(0)]\rangle \tag{6.13}$$

where $\mathbf{b}(t)$ and $\mathbf{b}(0)$ are the vectors of nucleon position at time t and initial instant t = 0. The angular brackets denote the ensemble average over all vectors and molecules. In powders and liquids, the I_{inc} must be averaged over all possible directions of $\mathbf{Q}$. In addition, molecular oscillations and lattice vibrations contribute to scattering. In order to take into account the last contributions, Equation (6.13) must be multiplied by the Debye–Waller factor $\exp(-2W)=\exp\{-\langle(\mathbf{Q}\times\mathbf{u})^2\rangle\}$, where $\mathbf{u}$ is the vector of the nuclear shift from its equilibrium position. In accord with Equations (6.8) and (6.9), $\mathbf{u} = \mathbf{u}_{fluct}{}^{trans}(t) + \mathbf{u}_{libr}{}^{rot}(t)$, and $\mathbf{b}(t) = \mathbf{r}_0{}^{rot}(t)$. The Debye–Waller factor is neglected in this treatment.

In the EAJ model framework, the quantity $F[\mathbf{Q},\mathbf{b}(t)] = \exp\{\mathrm{i}\,\mathbf{Q}\,\mathbf{b}(t)\}$ is a continuous function of the angular variable $g(t)$

$$F[\mathbf{Q},\mathbf{b}(t)] \equiv \exp\{\mathrm{i}\,\mathbf{Q}\,\mathbf{b}(t)\} \equiv \exp\{\mathrm{i}\,\mathbf{Q}\,\mathbf{b}[b,g(t)]\} \tag{6.14}$$

Therefore, we can calculate the ISF by means of the following integration:

$$I_{\mathrm{inc}}(Q,t) = \frac{1}{8\pi^2} \oiint \exp[-\mathrm{i}\,\mathbf{Q}\,\mathbf{b}(t)]\,\exp[\mathrm{i}\,\mathbf{Q}\,\mathbf{b}(0)]\,W(g,t;\,g_0,t_0)\,\mathrm{d}g\,\mathrm{d}g_0 \tag{6.15}$$

where $W(g,t;\,g_0,t_0)$ is the conditional probability density that the molecular vector **b** is oriented within the limits of the unit solid angle around the direction g at an instant t, assuming that it was oriented within the limits of the unit solid angle around the direction g_0 at the initial instant $t_0 = 0$. The conditional probability density $W(g,t;\,g_0,t_0)$ relates to the unconditional probability density $W(g_0,t,g)$ by

$$W(g,t;\,g_0,t_0) = 4\pi\, W(g_0,t,g) \tag{6.16}$$

where 4π is the value of the full solid angle. By taking into account the probability density (2.29) deduced in the EAJ model framework and assuming $t_0 = 0$, the conditional probability density $W(g,t;\,g_0,t_0)$ is expressed by

$$W(g,t;\,g_0,0) = \sum_{\alpha,\beta,\nu} \frac{2\nu+1}{\chi_{\alpha E}}\, q_\alpha^{(\nu)}\, \Psi_{\alpha\beta}^{(\nu)}(g_0)^*\, \Psi_{\alpha\beta}^{(\nu)}(g)\, \exp\left(-\frac{t}{\tau_\alpha}\right) \tag{6.17}$$

6.3 Incoherent neutron scattering function adapted to the EAJ model

One derives the ISF by developing $F[g(t)] = \exp\{\mathrm{i}\mathbf{Qb}[b,g(t)]\}$, the plane-wave formula, in a series of unitary spherical tensor components $Y_m^{(l)}(g)$ [21–24]:

$$F(g) = \exp\{\mathrm{i}\mathbf{Qb}[b,g(t)]\} = 4\pi \sum_{l=0}^{\infty} \mathrm{i}^l j_l(Qb) \sum_{m=-l}^{l} Y_m^{(l)}(\Lambda)\, Y_m^{(l)}(g) \tag{6.18}$$

where $j_l(Qb)$ is the spherical Bessel function of order l; and Q and Λ are respectively the modulus and the polar coordinates of the momentum transfer vector **Q** in the LRF.

Performing mathematical procedures prescribed by Equation (6.15), with the help of Equations (1.7), (3.1), (6.17), and (6.18), we obtain [116, 117]

$$
\begin{aligned}
I_{\text{inc}}(\Lambda,t) = {} & 2\oiint \sum_{l'=0}^{\infty} \mathrm{i}^{l'} j_{l'}(Qb) \sum_{m'=-l'}^{l'} Y_{m'}^{(l')}(\Lambda) Y_{m'}^{(l')}(g_0) \sum_{l=0}^{\infty} (-\mathrm{i})^{l} j_{l}(Qb)^{*} \\
& \cdot \sum_{m=-l}^{l} Y_{m}^{(l)}(\Lambda)^{*}\; Y_{m}^{*(l)}(g)^{*} \sum_{\alpha\beta\nu} \frac{2\nu+1}{\chi_{\alpha E}} q_{\alpha}^{(\nu)} \\
& \cdot \sum_{s'n'} \sum_{sn} D_{s'n'}^{(\nu)}(\Omega)^{*}\; D_{sn}^{(\nu)}(\Omega) \psi_{\alpha\beta s'}^{(\nu)*}\; \psi_{\alpha\beta s}^{(\nu)}\, Y_{n'}^{(\nu)}(g_0)^{*}\, Y_{n}^{(\nu)}(g) \exp\left(-\frac{t}{\tau_{\alpha}}\right) \\
& \cdot \mathrm{d}g\mathrm{d}g_0 \\
= {} & 2\sum_{\alpha\beta\nu} \frac{2\nu+1}{\chi_{\alpha E}} q_{\alpha}^{(\nu)} \sum_{l'=0}^{\infty} \mathrm{i}^{l'} j_{l'}(Qb) \sum_{m'=-l'}^{l'} \sum_{s'n'} Y_{m'}^{(l')}(\Lambda) D_{s'n'}^{(\nu)}(\Omega) \psi_{\alpha\beta s'}^{(\nu)} \\
& \cdot \int Y_{m'}^{(l')}(g_0) Y_{n'}^{(\nu)}(g_0)^{*} \mathrm{d}g_0 \sum_{l=0}^{\infty} (-\mathrm{i})^{l} j_{l}\,(Qb) \\
& \cdot \sum_{m=-l}^{l} \sum_{sn} Y_{m}^{(l)}(\Lambda)^{*} D_{sn}^{(\nu)}(\Omega)^{*} \psi_{\alpha\beta s}^{(\nu)*} \int Y_{m}^{(l)}(g)^{*}\; Y_{n}^{(\nu)}(g)\, \mathrm{d}g \exp\left(-\frac{t}{\tau_{\alpha}}\right) \\
= {} & 2\sum_{\alpha\beta\nu} \frac{2\nu+1}{\chi_{\alpha E}} q_{\alpha}^{(\nu)}\, \mathrm{i}^{\nu} j_{\nu}(Qb) \sum_{m'=-\nu}^{\nu} Y_{m'}^{(\nu)}(\Lambda) \sum_{s'} D_{s'm'}^{(\nu)}(\Omega) \psi_{\alpha\beta s'}^{(\nu)}\, \delta_{m'n'}\, \delta_{\nu l'} \\
& \cdot (-\mathrm{i})^{\nu} j_{\nu}(Qb) \sum_{m=-\nu}^{\nu} Y_{m}^{(\nu)}(\Lambda)^{*} \sum_{s} D_{sm}^{(\nu)}(\Omega)^{*}\; \psi_{\alpha\beta s}^{(\nu)*}\, \delta_{mn} \delta_{\nu l} \exp\left(-\frac{t}{\tau_{\alpha}}\right)
\end{aligned}
\tag{6.19}
$$

Applying the normalization conditions, the last expression reduces to

$$
I_{\text{inc}}(Q,t) = 2\sum_{\alpha\beta\nu} \frac{2\nu+1}{\chi_{\alpha E}} q_{\alpha}^{(\nu)} j_{\nu}^{2}(Qb) \left| \sum_{m,s=-\nu}^{\nu} \psi_{\alpha\beta s}^{(\nu)} D_{sm}^{(\nu)}(\Omega)\; Y_{m}^{(\nu)}(\Lambda) \right|^{2} \exp\left(-\frac{t}{\tau_{\alpha}}\right) \tag{6.20}
$$

where the three-dimensional angle Ω determines the orientation of the crystallographic reference frame (CRF) in the LRF. Taking into consideration the transformation rule, given by Equation (3.2), the single-crystalline expression of ISF takes the form

$$I_{inc}(Q,t) = 2\sum_{\alpha\beta\nu} \frac{2\nu+1}{\chi_{\alpha E}} q_{\alpha}^{(\nu)} j_{\nu}^{2}(Qb) \left|\Psi_{\alpha\beta}^{(\nu)}(\Lambda')\right|^{2} \exp\left(-\frac{t}{\tau_{\alpha}}\right) \tag{6.21}$$

where the functions $\Psi_{\alpha\beta}^{(\nu)}(\Lambda')$ are the basis functions of the HMM point symmetry group. The argument $\Lambda' \equiv (\Theta, \Phi)$ determines the direction of the momentum transfer vector **Q** in the CRF given by Θ and Φ, the polar angles of vector **Q** in the CRF.

The INSF consists of the sum of two contributions:

$$S_{inc} = S_{inc}(\text{elast}) + S_{inc}(\text{q-elast}) \tag{6.22}$$

One of them, $S_{inc}(\text{elast})$, the function of elastic incoherent neutron scattering, function (elastic INSF), does not depend on the energy transfer to the molecular system. The other, $S_{inc}(\text{q-elast})$ is called the function of quasi-elastic incoherent neutron scattering, (q-elastic INSF), by means of which the spectral distribution law of the energy exchange between the neutron beam and the molecules is described. Performing Fourier transform of Equation (6.21), we obtain the functions $S_{inc}(\text{elast})$ and $S_{inc}(\text{q-elast})$ in single-crystalline samples [116, 117]:

$$S_{inc}(\text{elast}) = 2\sum_{\nu=0}^{\infty}\sum_{\alpha}(2\nu+1) q_{\alpha=0}^{(\nu)} j_{\nu}^{2}(Qb) \left|\Psi_{\alpha=0}^{(\nu)}(\Lambda')\right|^{2} \delta(\omega) \tag{6.23}$$

and

$$S_{inc}(\text{q-elast}) = \frac{2}{\pi}\sum_{\nu=1}^{\infty}\sum_{\alpha}\sum_{\beta} \frac{(2\nu+1)}{\chi_{\alpha E}} q_{\alpha\neq 0}^{(\nu)} j_{\nu}^{2}(Qb) \left|\Psi_{\alpha\neq 0,\beta}^{(\nu)}(\Lambda')\right|^{2} \frac{\tau_{\alpha}}{1+\omega^{2}\tau_{\alpha}^{2}} \tag{6.24}$$

where $\delta(\omega)$ is the Dirac delta function and the label $\alpha = 0$ denotes the identical representation. In a powder, Equations (6.23) and (6.24) reduce to

$$S_{inc}(\text{elast}) = \frac{1}{2\pi}\sum_{\nu=0}^{\infty}\sum_{\alpha}(2\nu+1) q_{\alpha=0}^{(\nu)} j_{\nu}^{2}(Qb)\delta(\omega) \tag{6.25}$$

and

$$S_{inc}(\text{q-elast}) = \frac{1}{2\pi^{2}}\sum_{\nu=1}^{\infty}\sum_{\alpha}(2\nu+1)\, q_{\alpha\neq 0}^{(\nu)}\, j_{\nu}^{2}(Qb) \frac{\tau_{\alpha}}{1+\omega^{2}\tau_{\alpha}^{2}} \tag{6.26}$$

6.4 Discussion and comparison with the experiment

6.4.1 Theoretical outcomes

By means of the theory of incoherent neutron scattering in molecular crystals developed within the framework of the EAJ model, we established an unambiguous relation between the physical nature of various contributions to scattering and the symmetry properties of HMM. The hindered states symmetrized on the identical, irreducible representations give rise to the elastic scattering. The quasi-elastic scattering is determined by the hindered states symmetrized on the nonidentical, irreducible representations.

In single crystals, both theoretical INSF contributions are anisotropic. The graphs of the elastic INSF $S_{inc}(\text{elast}) = S_{inc}^{elast}$ and the peak values of the q-elastic INSF $S_{inc}(\text{q-elast}) = S_{inc}^{q\text{-}elast}$ in single crystals drawn according to Equation (6.23) and Equation (6.24) for the HMM symmetry groups O and T are shown in Figure 6.3(a) through Figure 6.3(d), and for the

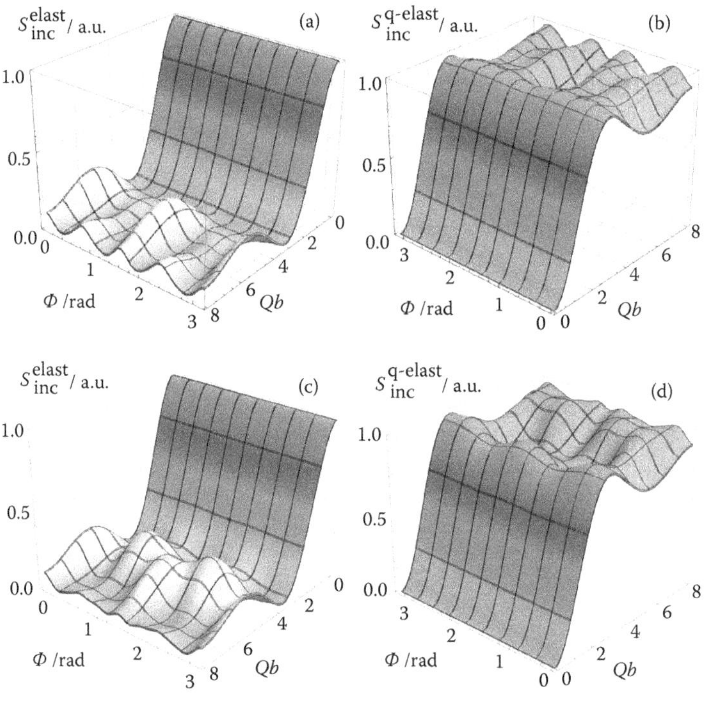

Figure 6.3 Surface plot graphs of the elastic INSF S_{inc}^{elast} and the peak values of the q-elastic INSF $S_{inc}^{q\text{-}elast}$ adapted to the HMM point symmetry groups O (a, b) and T (c, d).

HMM symmetry groups C_3 and C_2 in Figure 6.4(a) through Figure 6.4(d). They are presented in arbitrary units as functions of the angle Φ or Θ and the product *Qb*, where Φ is the azimuth angle, Θ is the polar angle, *Q* is the modulus of momentum transfer vector, and *b* is the modulus of the radius vector of scattering nucleus in the reference frame of HMM symmetry. For symmetry groups O and T of the cubic crystal system, the polar angle is taken equal $\Theta = 90°$. The graphs are fitted for values of the modulus of the momentum transfer vector satisfying the condition $0 \le Q \le 8/b$. Dynamical weights of the hindered states are taken equal to their static values.

The graphs of Figure 6.4 show more developed anisotropy of the scattering intensity for the HMM adapted to the groups of axial symmetry C_3 and C_2. The elastic scattering intensity being equal to unity has no dependence on the modulus of wave-vector transfer *Q* for the HMM symmetry axis directed along the transfer vector ($\Theta = 0$). The quasi-elastic

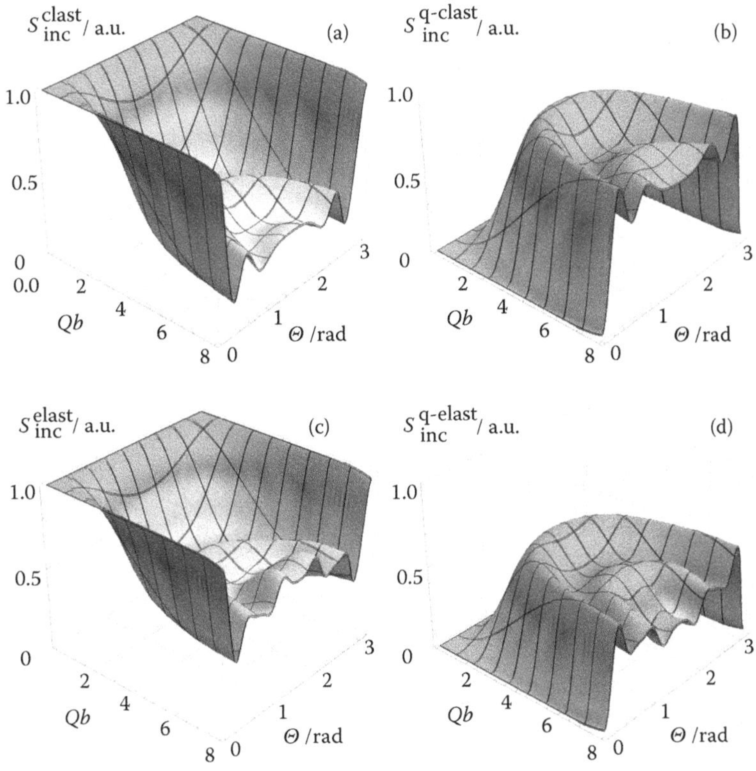

Figure 6.4 Surface plot graphs of the elastic INSF S_{inc}^{elast} and the peak values of the q-elastic INSF $S_{inc}^{q\text{-}elast}$ adapted to the HMM point symmetry groups C_3 (a, b) and C_2 (c, d).

scattering peak is equal to zero in this direction. This means that there is no energy transferring from the neutron beam to matter, and the incoherent scattering is completely elastic. For cubic symmetry motion groups, the anisotropy is not sharply expressed in both scattering intensities.

To throw more light on these outcomes, we present the elastic INSF S_{inc}(elast) for the HMM adapted to the octahedral symmetry group in analytical form:

$$S_{inc}(\text{elast}) = \left[\frac{1}{4\pi^2} j_0^2(Qb) + \frac{3}{8\pi} q_0^{(4)} j_4^2(Qb) \left| \sqrt{14}\, Y_0^{(4)}(\Theta,\Phi) \right.\right.$$
$$\left. + \sqrt{5}\left[Y_{-4}^{(4)}(\Theta,\Phi) + Y_4^{(4)}(\Theta,\Phi) \right] \right|^2$$
$$+ \frac{13}{16\pi} q_0^{(6)} j_6^2(Qb) \left| \sqrt{2}\, Y_0^{(6)}(\Theta,\Phi) - \sqrt{7}\left[Y_{-4}^{(6)}(\Theta,\Phi) + Y_4^{(6)}(\Theta,\Phi) \right] \right|^2$$
$$+ \frac{17}{384\pi} q_0^{(8)} j_8^2(Qb) \left\{ \left| 3\sqrt{22}\, Y_0^{(8)}(\Theta,\Phi) + 2\sqrt{7}\left[Y_{-4}^{(8)}(\Theta,\Phi) + Y_4^{(8)}(\Theta,\Phi) \right] \right.\right.$$
$$\left.\left.\left. + \sqrt{65}\left[Y_{-8}^{(8)}(\Theta,\Phi) + Y_8^{(8)}(\Theta,\Phi) \right] \right|^2 \right\} + \cdots \right] \delta(\omega) \qquad (6.27)$$

where $Y_m^{(\nu)}(\Theta,\Phi)$ is the component of a spherical harmonic, $j_\nu(Qb)$ is a first-order spherical Bessel function, and Θ and Φ are the polar angles of the vector **Q**.

The theoretical relative intensities of elastic INSF S_{inc} (elast) and q-elast INSF S_{inc}(q-elast) for powders are described by Equations (6.25) and (6.26). An example of the whole theoretical spectrum of the full incoherent neutron scattering $S_{inc}(Qb,\omega\tau)$ is shown in Figure 6.5(a). The graph is drawn in arbitrary units according to Equations (6.22), (6.25), and (6.26). The symmetry group of nuclear vector hindered motion is presented by the point group C_3. Dynamical weights of the hindered states are taken equal to their static values. The spectrum of quasi-elastic scattering intensity is presented by one Lorentzian line and expressed as a function of the dimensionless quantities $\omega\tau$ and Qb, where τ and b are fixed constants. It is seen from Figure 6.5(a) that the elastic contribution to the scattering function (the central curve, which corresponds to $\omega\tau = 0$) does not decrease down to zero by increasing the modulus of momentum transfer vector. By comparison, the elastic part of scattering intensity decreases to zero in the framework of the RD model [Figure 6.5(b)].

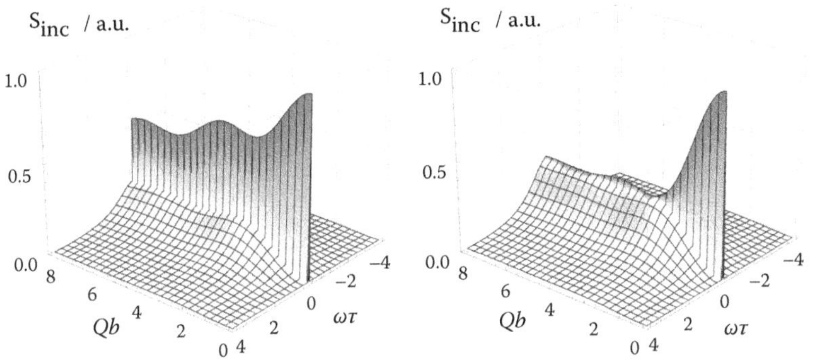

Figure 6.5 Incoherent neutron scattering function S_{inc} adapted to the HMM of the point symmetry group C_3 (a) and the continuous symmetry group O^{3+} (b). Elastic and quasi-elastic components of scattering intensity are drawn in arbitrary units by using Eqs (6.25) and (6.26) as functions of Qb and $\omega\tau$: Q is the modulus of momentum transfer vector ($0 \le Q \le 8/b$), b is the modulus of the radius vector of the scattering nucleus, ω is the scattering energy in frequency units and τ is the average correlation time of HMM.

In Figure 6.6, the two-dimensional graphs of the elastic INSF $S_{inc}(\text{elast}) = S_{inc}^{elast}$ are shown as a function of the dimensionless quantity Qb for various rotation symmetry groups in powders: 1, O^{3+}; 2, O; 3, T; 4, C_6; 5, C_4; 6, C_3; and 7, C_2. The curves labeled by the numbers 1, 2, and 3 are placed closely to each other for values $0 < Qb < 3$. The curves 4, 5, 6, and 7 are also close to each other for values $0 < Qb < 2$, but they are spaced considerably from curves 1–3. Curve 1, drawn for the square of the function $j_0^2(Qb)$, describes the elastic scattering in the framework of isotropic HMM (that is, the RD model).

In the experimental applications, the explicit expressions of graphs 2–7 shown in Figure 6.6 might be important. Taking the dynamical weights $q_\alpha^{(\nu)}$ equal to their static values, we can present Equation (6.25) by the linear combination of the squares of Bessel spherical functions $j_\nu^2(Qb) = [j_\nu(Qb)]^2$:

$$S_{inc}(\text{elast}) = \frac{1}{2\pi}\sum_{\nu=0}^{\infty}\left(\sum_{\alpha}\chi_{\alpha E}\right) j_\nu^2(Qb)\,\delta_{\alpha,0}\,\delta\omega = \frac{1}{2\pi}\sum_{\nu=0}^{\infty} n_0^{(\nu)} j_\nu^2(Qb)\,\delta\omega \qquad (6.28)$$

where the factor $n_0^{(\nu)}$ expresses the number of identical representations, which appropriates to the reducible representation $D^{(\nu)}$ of the group O^{3+}. For the abstract point groups, these factors are collected in Table 6.3. It

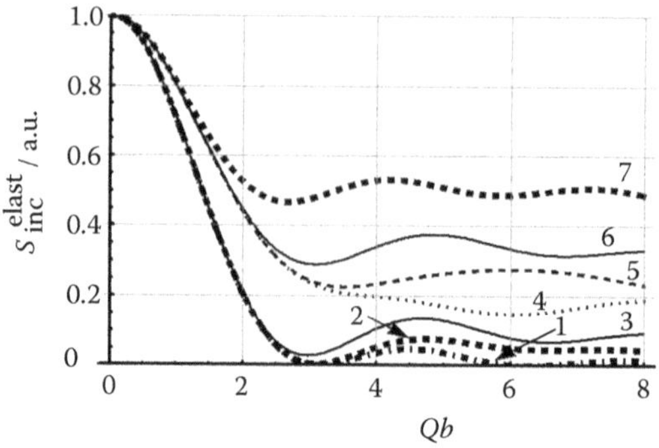

Figure 6.6 Graphs of the elastic incoherent neutron scattering functions $S_{inc}(\text{elast}) = S_{inc}^{elast}$ adapted to the HMM point symmetry groups O, 2; T, 3; C_6, 4; C_4,5; C_3, 6; and C_2, 7. Curve 1 corresponds to the function $S_{inc}(\text{elast})$ predicted in the RD-model framework for the isotropic HMM.

should be noted that formula (6.28) is the powder expression of elastic INSF. It is not difficult to surmise that the parallel expression of q-elastic INSF [Equation (6.26)] reduces for the HMM described by only one correlation time τ to

$$S_{inc}(\text{q-elast}) = \frac{1}{2\pi^2}\sum_{\nu=1}^{\infty}\left(2\nu+1-n_0^{(\nu)}\right)j_\nu^2(Qb)\frac{\tau}{1+\omega\tau} \tag{6.29}$$

Table 6.3 Number of Identical Irreducible Representations $n_0^{(\nu)}$ Contained in Some Crystallographic Point Symmetry Groups

	ν	$n_0^{(\nu)}$ 0	1	2	3	4	5	6	7	8
Point Groups	C_2	1	1	3	3	5	5	7	7	9
	C_3	1	1	1	3	3	3	5	5	5
	C_4	1	1	1	1	3	3	3	3	5
	C_6	1	1	1	1	1	1	3	3	3
	D_2	1	0	2	1	3	2	4	3	5
	D_3	1	1	1	1	2	1	3	2	3
	D_4	1	0	1	0	2	1	2	1	3
	D_6	1	0	1	0	1	0	2	1	2
	T	1	0	0	1	1	0	2	1	1
	0	1	0	0	0	1	0	1	0	1

6.4.2 Experimental applications

6.4.2.1 Rubidium hydrosulphide RbSH

As a first example of the application of the new expression for INSF, we discuss the experimental data measured in powder rubidium hydrosulphide, RbSH [118]. They are shown for elastic INSF by full ($T = 373$ K) and open ($T = 393$ K) circles in Figure 6.7 and labeled by the symbol *a*. This is the case when the intensity of elastic scattering does not decrease down to zero for large momentum transfer vectors and hence the RD model approach does not approximate the HMM.

The structure of RbSH (isomorphous with NaSH in both the high-temperature face-centered cubic (FCC) phase and intermediate trigonal phase shown in Figure 6.8) is identified with the structure of the trigonal distorted lattice NaCl [118, 119]. It is assumed that SH^- anions occupy the nodes of the cube and the chemical bond is directed along the diagonal axis [1,1,1]. The alignment of SH^- ions shown in the figure is that deduced for the trigonal phase of NaSH, with the ions aligned along the trigonal axis. This trigonal axis is closely related to a [1,1,1] axis in the high-temperature FCC phase. It should be noted that in the cubic phase

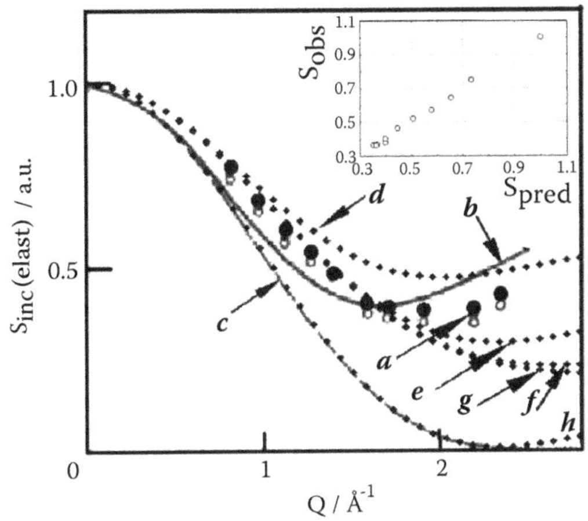

Figure 6.7 Graphs of experimental (*a*) and theoretical (*b*)–(*h*) functions of elastic incoherent neutron scattering in polycrystalline RbSH. Open and dark circles *a* are the experimental data in the temperature interval 373 K ≤ *T* ≤ 393 K [118]. Theoretical curves *b* and *c* are drawn in the framework of the FAJ-model approach for the proton vector jumps between two and eight equivalent orientations, respectively; and those denoted by *d, e, f, g,* and *h* are drawn in the framework of the EAJ-model approach for the HMM symmetry groups C_2, C_3, C_4, C_6, and O. With permission from Rowe, J. M., et al., *J. Chem. Phys.* 59, no. 12 (1973): 6652–6655.

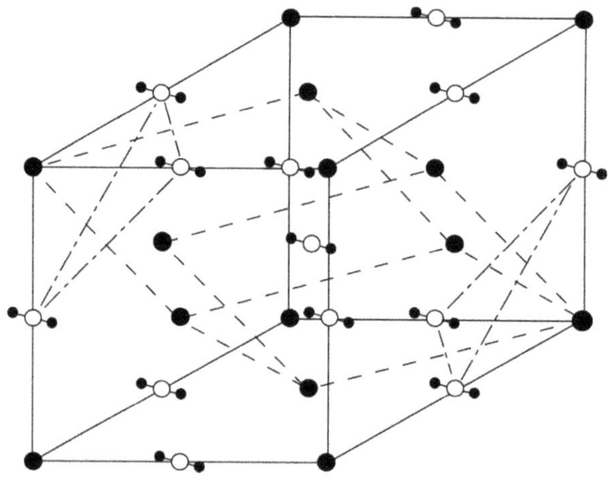

Figure 6.8 Composite drawing of the suggested RbSH crystal structures [118]. The SH^- ions are shown aligned along a single [111] direction of the high-temperature FCC (or low-temperature pseudo cubic) cell. One of these diagonals is shortened in the trigonal phase and the dashed lines outline the primitive lower temperature trigonal cell. The dot-dash lines outline triangles of near-neighbor SH^- ions in the (111) planes perpendicular to the trigonal axis. The S–H distance is $b^{SH} = 1.35 \pm 0.04$ Å (1 Å = 10^{-10} m); ● Rb; ○ S; • 1/2 H. With permission from Rowe, J. M., et al., *J. Chem. Phys.* 59, no. 12 (1973): 6652–6655.

the ions are randomly oriented along a set of crystal directions to assume the requisite cubic symmetry.

The former theoretical analysis of the proton-vector hindered rotation was made within the framework of the fixed angular jump (FAJ) model by suggesting that vector $\mathbf{b}_{SH}$ exhibits rotational jumps between 2 (curve *b*) and 8 (curve *c*) fixed sites (Figure 6.7). We can see that these theoretical curves describe roughly the experimental data. Curve *b* is placed more closely to the experimental points (*a*). Therefore, we accepted the model in which framework the proton vectors exhibit random jumps between two opposite equilibrium positions was accepted [118].

Curves *d* through *h*, calculated by using Equation (6.28) and the numerical data from Table 6.1, are presented in the same figure. They were drawn assuming that vector $\mathbf{b}_{SH}$ exhibits hindered motions adapted in the framework of EAJ model to the abstract point symmetry groups of pure rotation C_2 (curve *d*), C_3 (curve *e*), C_4 (curve *f*), C_6 (curve *g*), and tetrahedron T (curve *h*). The modulus of the vector $\mathbf{b}_{SH}$ was taken equal to the typical length of chemical bond S–H: $b_{SH} = 1.325$ Å (1 Å = 10^{-10} m). We can see that the curve *e* gives the best agreement to the experimental data.

With respect to the HMM theory developed in the framework of the EAJ model, factor $n_0^{(\nu)}$ can be considered as a fitted parameter in Equation (6.28). In conformity with the normalization condition $\Sigma_\alpha n_\alpha^{(\nu)} = 2\nu + 1$, it can take numerical values within the limits from 0 up to 2ν + 1. Extrapolating the experimental points *a* of Figure 6.7 by Equation (6.28) with the help of Newton's method gives the following expression:

$$S_{\text{inc}}(\text{elast}) \approx \frac{1}{2\pi}\Big[j_0^2(Qb) + 0.37\, j_1^2(Qb) + 1.45\, j_2^2(Qb) + 2.56\, j_3^2(Qb) + 7.96\, j_4^2(Qb) \Big]\delta\omega. \tag{6.30}$$

Because terms higher than the fourth rank give such a small contribution to the scattering intensity, they are omitted in the expansion (6.30). In the small window of Figure 6.7, the diagram of conformity between the experimental values of the elastic scattering function S_{observ} and those predicted by the extrapolated Equation (6.30) S_{predic} is shown.

Factor $n_0^{(\nu)}$, taken from fitted Equation (6.30) and those computed for abstract point symmetry groups C_2, C_3, and C_4 (Table 6.3), are placed in Table 6.4. It is now possible to establish an approximate geometrical symmetry group of the hindered motion of SH^- anions by comparing the respective values of $n_0^{(\nu)}$. We can see from the curves of Figure 6.7 and the data of Table 6.4 that the group C_3 gives better agreement than groups C_2 or C_4, which is consistent with the structure of a trigonal distorted RbSH-lattice structure of NaCl and the stable position of the SH vector oriented along the [1,1,1] direction. However, this outcome is not consistent with the original discussion of the experimental data [118].

6.4.2.2 *Zinc tetraamine perchlorate* $[Zn(NH_3)_4](ClO_4)_2$

Let us present another description of experimental study on the quasi-elastic incoherent neutron scattering in powder zinc tetraamine perchlorate $[Zn(NH_3)_4](ClO_4)_2$. The experimentally determined elastic incoherent

Table 6.4 Values of the Factors $n_0^{(\nu)}$ Taken from Table 6.3 and by Extrapolating Equation (6.30)

	$n_{\alpha=0}^{(\nu)}$				
Citation	ν = 0	ν = 1	ν = 2	ν = 3	ν = 4
Group C_2	1	1	3	3	5
Group C_3	1	1	1	3	3
Group C_4	1	1	1	1	3
Equation (6.30)	1	0.37	1.45	2.56	7.96

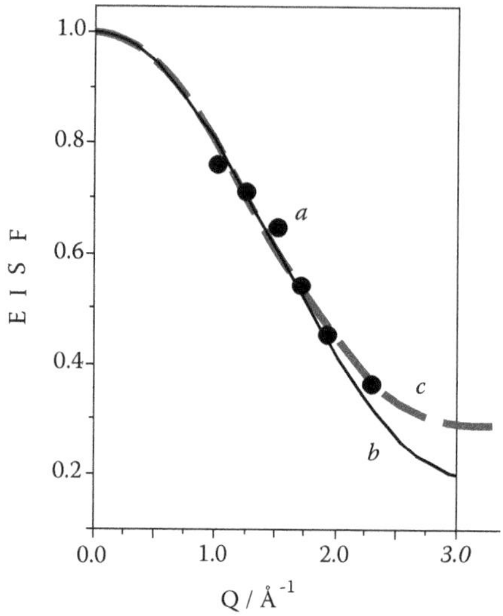

Figure 6.9 Elastic incoherent structure factor (EISF) determined experimentally in powder $[Zn(NH_3)_4](ClO_4)_2$ for T = 290 K (filled circles *a*) [33]. Theoretical lines drawn by using the FAJ model with NH vectors reorienting about three-fold axis (solid lines *b*) [33] and by using the EAJ model with NH vector reorientation symmetry group C_3 of trigonal crystallographic system (dashed lines *c*).

structure factor EISF (elastic INSF, S_{inc}^{elast}) is shown for six selected Q values at temperature 290 K by filled circles *a* in Figure 6.9 [33]. The discussion of these data has been performed in the framework of the FAJ model yielded by the instantaneous proton jumps to the angle 120° around the three-fold axis of rotational symmetry for NH_3 ligands. Solid line *b* presents the theoretical results calculated for the proton distance to the axis of rotation, b = 0.80 Å (1 Å = 10^{-10} m). We note that there is a good agreement between the experimental and theoretical results for the initial region of wave-vector transfer up to the value Q = 2 Å^{-1}. Our calculation of EISF by using the EAJ model for the HMM of NH vectors adapted to the point symmetry group C_3 is pictured by dashed lines *c*. It is seen from Figure 6.9 that the dashed line better approximates the experimental points.

6.4.2.3 *Ammonium zinc trifluoride* NH_4ZnF_3

The room-temperature structure of ammonium zinc trifluoride (zincate-trifluoro-ammonium) (NH_4ZnF_3) is of the perovskite type, space group *Fm3m*, which is shown in Figure 6.10 with the tetrahedral NH_4^+ ion in the center of the cell and constant a_0 = 4.1181(2) Å for the cubic cell at room temperature [120]. This crystal exhibits a structural transition to tetragonal phase at T_C = 115 K so that just below T_C the tetragonal lattice parameters are a_0 = 4.0822(4) Å and c_0 = 4.1465(4) Å.

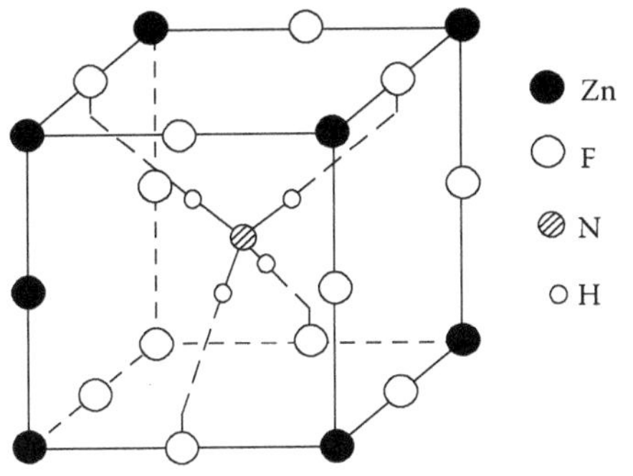

Figure 6.10 Crystallographic structure of perovskite NH_4ZnF_3 in the cubic phase: NH_4 atomic group is shown as a regular tetrahedron [120]. With permission from Steenbergen, Chr., et al., *J. Chem. Phys.* 70, no. 3 (1979): 1450–1455.

As regards the geometry of the NH_4^+ atomic group, Figure 6.10 shows the position of a purely tetrahedral NH_4^+ ion, in which the H atoms have a little shift from the lines connecting the N atom at the center to the neighboring fluorine ions at the mid-edge positions.

The quasi-elastic neutron scattering (QENS) experiments were performed to study the dynamics and the geometry of NH_4^+ ion motion. The lack of single crystals was compensated partially by the use of large momentum transfers, where the elastic intensity is most sensitive to the site geometry of the NH_4^+ ions. At room temperatures, it was suggested that the regular ammonium ion exhibits random rotation jumps between 12 equivalent equilibrium positions [120]. We shall call such model the FAJ model with 12 sites. By lowering the temperature, the cubic unit cell of NH_4ZnF_3 changes to that tetragonal followed a little deformation (distortion) of the ammonium tetrahedron along the *c* axis of the unit cell. As a result, the NH_4^+ ion modifies its dynamics to the model of discrete jumps (FAJ model) among the four positions around the cell axis [120]. In such cases where the ion distortion does not manifest, the model of ion motion will be called the FAJ model for the undistorted NH_4^+ ion with four sites. Accounting for the ion distortion in the NH_4^+ ion motion allows one to call that motion the FAJ model for the distorted NH_4^+ ion with four sites.

Because the aim of this book is to demonstrate the capacity of the EAJ model, we shall interpret the elastic incoherent scattering data, which reflect mainly the crystal structure. The experimental and theoretical data for elastic incoherent structure factor (EISF) at three temperatures (199.5 K, 228 K, and 273 K) are shown in Figure 6.11 [120]. The various theoretical curves, which approximate the experimental points, are also shown: 1, FAJ model with 12 sites; 2, FAJ model for the undistorted NH_4^+ ion with four sites; 3, FAJ model for the distorted NH_4^+ ion with four sites; 4, EAJ

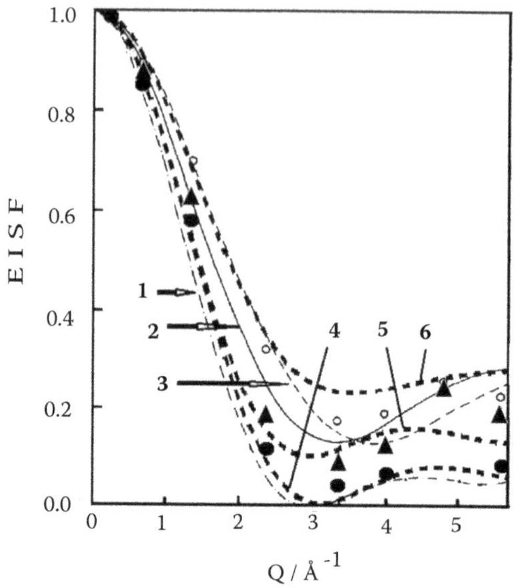

Figure 6.11 Elastic incoherent structure factor (EISF) S_{inc}(elast) for three temperatures: O, T = 199.5 K; ▲, T = 228 K; and •, T = 273 K. The various theoretical curves that approximate the experimental points are also shown: 1, FAJ model with 12 sites; 2, FAJ model for the undistorted NH_4^+-ion with four sites; 3, FAJ model for the distorted NH_4^+ ion with four sites; 4, EAJ model adapted to the point symmetry group of octahedron O; 5 and 6, EAJ model adapted to the respective point symmetry groups D_4 and C_4 of tetragonal crystal system. With permission from Steenbergen, Chr., et al., *J. Chem. Phys.* 70, no. 3 (1979): 1450–1455.

model adapted to the point symmetry group of octahedron O; 5 and 6, EAJ model adapted respectively to the point symmetry groups D_4 and C_4 of a tetragonal crystal system.

The present description of EISF for the crystal NH_4ZnF_3 within the limits of the EAJ model indicate with a high degree of assurance that at room temperature the hindered motion of ammonium ions takes place by means of random ion rotation adapted to the octahedral symmetry. Our interpretation of the QENS experimental data (curves 5 and 6 of Figure 6.11) suggests that during cooling the structure transition from a cubic to a tetragonal crystallographic system forces a lowering of the symmetry of HMM from octahedral to a point group of tetragonal crystal system D_4 or C_4.

In conclusion, we emphasize some advantages of the description of the relative intensity of incoherent neutron scattering in the framework of the EAJ model taken as the HMM model.

- Studying the scattering in single crystals lets us obtain quantitative knowledge of the symmetry of the molecule motion and the symmetry of its site.
- The elastic part of the INSF is symmetrized in the identical, irreducible representations; the quasi-elastic part is symmetrized in the nonidentical, irreducible representations of the HMM symmetry group.
- The scattering is anisotropic in all molecular single crystals. In the cases of HMM adapted to the cubic symmetry groups, this anisotropy is feebly expressed.
- The residual elastic scattering justified by symmetry is prescribed for large values of momentum transfer vector in solids.

References

1. Nye, J. F. *Physical Properties of Crystals*. New York: Oxford University Press, 1985.
2. Kitaigorodsky, A. I. *Molecular Crystals and Molecules*. New York: Academic Press, 1973.
3. Sirotin, Yu. I. and Shaskolskaia, M. P. *Fundamentals of Crystal Physics*. Moscow: Nauka, 1979.
4. Gutowsky, H. S. and Pake, G. E. "Structural Investigations by Means of Nuclear Magnetism. II. Hindered Rotation in Solids." *J. Chem. Phys.* 18 (1950): 162–169.
5. Bashirov, F. and Gaisin N. "The Theory of Hindered Molecular Motion and Its Application to Spectroscopic Studies." *Crystallogr. Rev.* 16 (2010): 3–87.
6. Debye, P. *Polar Molecules*. New York: Reinhold, 1929.
7. Kittel, Ch. *Introduction to Solid-State Physics*. New York: Wiley, 2004.
8. Brown, W. F. *Dielectrics*. Leipzig: Springer-Verlag, 1956.
9. Coffey, W., Evans, M., and Grigolini, P. *Molecular Diffusion and Spectra*. New York: Wiley, 1984.
10. Rakov, A. V. "Studying Brownian Rotation Motion of Molecules by the Methods of Raman Scattering and Infrared Absorption in Condensed Matter." *Proc. Lebedev's Phys. Inst.: Res. Mol. Spect.* 27 (1964): 111–149.
11. Gordon, R. G. "Molecular Motion in Infrared and Raman Spectra." *Chem. Phys.* 43, no. 4 (1965): 1307–1312.
12. Anderson, A. *The Raman Effect*. New York: Marcel Dekker, 1973.
13. Williams, G. "Time-Correlation Functions and Molecular Motion." *Chem. Soc. Rev.* 7, no. 7 (1978): 89–131.
14. Long, D. A. *The Raman Effect. A Unified Treatment of the Theory of Raman Scattering by Molecules*. New York: Wiley, 2002.
15. Valiev, K. A. *Research of Liquid Matter by Spectroscopy Techniques*. Moscow: Nauka, 2005.
16. Bloembergen, N. "Processes of Spin Relaxation in Two-Proton System." *Phys. Rev.* 104 (1956): 1542–1548.
17. Abragam, A. *The Principles of Nuclear Magnetism*. New York: Oxford University Press, 1961.
18. Slichter, Ch. P. *Principles of Magnetic Resonance*. New York: Harper and Row, 1965.
19. Hertz, H. G. "The Problem of Intramolecular Rotation in Liquids and Nuclear Magnetic Relaxation." *Prog. NMR Spect.* 16 (1983): 115–162.

20. Squires, G. L. *Introduction to the Theory of Thermal Neutron Scattering*. New York: Dover, 1996.
21. Willis, B. T. M. *Thermal Neutron Scattering*. Oxford: Oxford University Press, 1973.
22. Lovesey, S. W. and Springer T. *Dynamics of Solids and Liquids by Neutron Scattering*. Berlin: Springer-Verlag, 1977.
23. Celotta, R. and Levine, J., ed. *Methods of Experimental Physics*. Vol. 23, *Neutron Scattering*, ed. by Skold, K. and Price, D.L. Orlando, Fla.: Academic Press, 1986.
24. Bée, M. *Quasi-Elastic Neutron Scattering Principles and Applications in Solid-State Chemistry, Biology and Material Science*. Bristol: Adam Hilger, 1988.
25. Tomita, K. "States of Solid Methane as Inferred from Nuclear Magnetic Resonance." *Phys. Rev.* 89, no. 2 (1953): 429–438.
26. Gordon, R. G. "Correlation Functions for Molecular Motion." *Adv. Magn. Reson.* 3 (1968): 1–42.
27. Bashirov, F. I., Popov, Yu. L., Saikin, K. S., and Dautov, R. A. "Nuclear Magnetic Relaxation Induced by Random Reorientations of Molecules in Crystals." *Zh. Eksp. Teor. Fiz.* 62 (1972): 1803–1810. (Engl. Transl.: *Sov. Phys. JETP* 35 [1972]: 937–940).
28. Rigny, P. "Réorientations dans les Cristaux Moléculaires et Fonctions de Corrélation." *Physica* 59 (1972): 707–721.
29. Valiev, K. A. and Ivanov, E. N. "Rotational Brownian Motion." *Uspehi Fiz. Nauk.* 109 (1973): 31–64.
30. Haeberlen, U. *High Resolution NMR in Solids: Selective Averaging*, ed. Waugh, J. S. New York: Academic Press, 1976.
31. Holderna-Natkaniec, K., Jurga, K., Natkaniec, I., Nowak, D., and Szyczewski, A. "Molecular Dynamics of Ethisterone Studied by ^{1}H NMR, IINS and Quantum Mechanical Calculations." *Chem. Phys.* 317 (2005): 178–187.
32. Szyczewski, A. and Hołderna-Natkaniec, K. "Molecular Dynamics of 17a- and 21-Hydroxy Progesterone Studied by NMR: Relation between Molecule Conformation and Height of the Barrier for Methyl Group Reorientations in Steroid Compounds." *J. Molec. Struct.* 734 (2005): 129–136.
33. Migdał-Mikuli, A., Hołderna-Natkaniec, K., Mikuli, E., Hetmanŕczyk, Ł., and Natkaniec, I. "Phase Transitions and NH_3 Motions in $[Zn(NH_3)_4](ClO_4)_2$ Studied by Incoherent Neutron Scattering and ^{1}H NMR Methods." *Chem. Phys.* 335 (2007): 187–193.
34. Hetmanczyk, J., Migdał-Mikuli, A., Mikuli, E., Hołderna-Natkaniec, K., Hetmanŕczyk, Ł., and Natkaniec, I. "Phase Transitions and H_2O Motions in $[Ca(H2O)_4](ClO_4)_2$ Studied by Infrared Spectroscopy, Inelastic/Quasi-Elastic Incoherent Neutron Scattering and Proton Magnetic Resonance. Part II." *J. Mol. Struct.* 923 (2009): 103–109.
35. Bashirov, F. I. *Symmetrized Angular Autocorrelation Functions*. Extended abstracts of XXVIIth Congress AMPERE, Magnetic Resonance and Related Phenomena, Vol. 1. Kazan: KFTI-Kazan, (1994): 282–283.
36. Bashirov, F. I. "Angular Autocorrelation Functions in Molecular Crystals: Application to NMR-Relaxation and Raman Spectra." *Mol. Phys.* 91, no. 2 (1997): 281–300.
37. Feller, W. *An Introduction to the Probability Theory and Its* Application. Princeton, NJ: Princeton University Press, 1968.
38. Wigner, E. *Gruppentheorie*. Berlin: Braunschweig, 1931.

39. Landau, L. D. and Lifshitz, E. M. *Quantum* Mechanics. Moscow: Fizmatgiz, 1963.
40. Valiev, K. A. and Eskin, L. D. "On the Rotational Diffusion of Molecules and Light Scattering in Liquids." *Optika i Spektroskopia* 12 (1962): 758–764.
41. Ivanov, E. N. "Theory of Rotational Brownian Motion." *Zh. Eksp. Teor. Fiz.* 45 (1963): 1509–1517. (Engl. Transl.: *Sov. Phys. JETP* 18 [1964]: 1041–1045).
42. Sidorova, A. I., Batisheva, M. G., and Shermatov, E. H. "Temperature Broadening of Infrared Absorption Band of Acetonitrile and Its Relation to Rotational Mobility of Molecules." *Optica i Spectroscopia: Mol. Spect.* 2 (1963): 188–191.
43. Wall, T. T. "Time Correlations from Infrared Bands of HDO." *J. Chem. Phys.* 52, no. 5 (1969): 2792–2793.
44. Valiev, K. A. "To the Theory of Line Width of Vibrational Spectra of Molecules in Liquids. Effect of Rotation of Molecules on the Line Width of Infrared Absorption." *Optika i Spektroskopia: Molecularnaia Spektroskopia* 2 (1963): 98–103.
45. Wigner, E. "Symmetry and Conservation Laws." *Phys. Today* 92, no. 17 (1964): 34–45.
46. Zimpel, Z. and Medycki W. "Theory of the Effect of Random Rotational Jumps on the Nuclear Spin-Lattice Relaxation in Solids." *J. Magn. Reson.* 92 (1991): 377–397.
47. Kodama, T. "Proton Spin-Lattice Relaxation and Order-Disorder Transition in Ammonium Chloride." *J. Magn. Reson.* 7 (1972): 137–160.
48. Michel, K. H. "Spin-Lattice Relaxation and Molecular Reorientations near T_C with Application to NH_4Cl." *J. Chem. Phys.* 58 (1973): 142–152.
49. Bashirov, F. I. "Nuclear Magnetic Relaxation and Reorientation of Molecules in Crystals." PhD thesis, Kazan State University, Kazan, 1972.
50. Ivanov, E. N. "Theory of Nuclear Spin-Lattice Relaxation in Molecular Crystals." *Fiz. Tverd. Tela* 17 (1975): 851–858.
51. Tang, J., Sterna, L., and Pines, A. "Anisotropic Spin-Lattice Relaxation of Deuterated Hexamethylbenzene." *J. Magn. Reson.* 41 (1980): 389–394.
52. Watton, A., Sharp, A. R., Petch, H. E., and Pintar, M. M. "Proton Magnetic Resonance Study of the Spin-Symmetry States of Ammonium Ions in Solids." *Phys. Rev. B* 5 (1972): 4281–4291.
53. Watton, A. "Nuclear Spin-Lattice Relaxation from Hindered Rotations in Dipolar Solids." *Phys. Rev. B* 17 (1978): 945-951.
54. O'Reilly, D. E. and Tsang, T. "Deuteron Magnetic Resonance and Proton Relaxation Times in Ferroelectric Ammonium Sulfate." *J. Chem. Phys.* 46, no. 4 (1967): 1291–1300.
55. Bildanov, M. M., Zaripov, M. R., and Andreev, N. K. "Effect of Molecule Symmetry to the Rate of Magnetic Relaxation of Nuclei in Solids." *Phiz. Tver. Tela* 15 (1973): 2253–2255.
56. Amoureux, J. P., Bée, M., and Virlet, J. "Anisotropic Molecular Reorientations of Adamantane in its Plastic Solid Phase: ^{1}H NMR Relaxation Study in Solid Solutions of $C_{10}H_{16}$ and $C_{10}D_{16}$." *Mol. Phys.* 41 no. 2 (1980): 313–324.
57. Bée, M., Amoureux, J. P., and Lechner, R. E. "Incoherent Quasi-Elastic Neutron Scattering Study of Molecular Motion in 1-Cyanoadamantane." *Mol. Phys.* 41, no. 2 (1980): 325–329.
58. Low, W. *Paramagnetic Resonance in Solids.* New York: Academic Press, 1960.

59. Halford, R. S. J. "Motion of Molecules in Condensed Systems: I. Selection Rules, Relative Intensities, and Orientation Effects for Raman and Infrared Spectra." *Chem. Phys.* 14, no. 1 (1946): 8–15.
60. Bhagavantam, S. and Venkataryudu, T. *Theory of Groups and Its Application to Physical Problems*. New York: Academic Press, 1969.
61. Ballhausen, C. J. *Introduction to Ligand-Field Theory*. New York: McGraw-Hill, 1962.
62. Hornig, D. F. "The Vibrational Spectra of Molecules and Complex Ions in Crystals. I. General Theory." *J. Chem. Phys.* 16, no. 11 (1948): 1063–1076.
63. Leach, A. R. *Molecular Modelling: Principles and Applications*. Harlow: Prentice Hall (Pearson Education), 2001.
64. Bashirov, F. I. "Spectroscopy of the Hindered Molecular Motion in Condensed Molecular Media." *Asian J. Spect.* 4 (2000): 97–117.
65. Knox, R. S. and Gold, A. *Symmetry in the Solid State*. New York: Amsterdam: W. A. Benjamin, 1964.
66. Petrashen, M. I. and Trifonov, E. D. *Application of Group Theory to Quantum Mechanics*. Moscow: Nauka, 1967.
67. Streitwolf, H. *Gruppentheorie in der festkorperphysik*. Leipzig: Akademische Verlaggesellschaft, 1967.
68. Leushin, A. M. *Tables of Functions Transforming According to the Irreducible Representation of the Crystallographic Point Groups*. Moscow: Nauka, 1968.
69. Bashirov, F. I. "Dielectric Properties Induced by Hindered Molecular Motion in Crystals and Liquids." *Eur. Phys. J. Appl. Phys.* 8 (1999): 99–104.
70. Bashirov, F. I. and Gaisin, N. K. "Dielectric Spectra in Molecular Single Crystals." *Centr. Euro. J. Phys.* 7 (2009): 79–83.
71. Bartel, J., Bachhuber, K., Buchner, R., and Hetzenaner H. "Dielectric Spectra of Some Common Solvents in the Microwave Region. Water and Lower Alcohols." *Chem. Phys. Lett.* 165 (1990): 369–373.
72. Bashirov, F. I. and Gaisin, N. K. "Shape of Molecular Infrared Absorption and Raman Scattering Lines as Probe of Hindered Molecular Motion and Site Symmetry in Crystals." *J. Ram. Spect.* 29 (1998): 131–142.
73. Huntress, W. T., Jr. "Effects of Anisotropic Molecular Rotational Diffusion on Nuclear Magnetic Relaxation in Liquids." *J. Chem. Phys.* 48, no. 8 (1968): 3524–3533.
74. Miller, R. E., Getty, R. R., Treuil, K. L., and Leroi, G. E. "Raman Spectrum of Crystalline Lithium Nitrate." *J. Chem. Phys.* 51, no. 4 (1969): 1385–1389.
75. Rousseau, D. I., Miller, R. E., and Leroi, G. E. "Raman Spectrum of Crystalline Sodium Nitrate." *J. Chem. Phys.* 48, no. 8 (1968): 3409–3413.
76. Dawson, P., Hargreave, M., and Wilkinson, G. R. "The Vibrational Spectrum of Zircon." *J. Physics C* 4 (1971): 240–249.
77. Hubbard, P. S. "Nonexponential Nuclear Magnetic Relaxation by Quadrupole Interactions." *J. Chem. Phys.* 53, no. 3 (1970): 985–987.
78. Bloembergen, N., Purcell, E. M., and Pound, R. V. "Relaxation Effect in Nuclear Magnetic Resonance Absorption." *Phys. Rev.* 73 (1948): 679–712.
79. Gutowsky, H. S., Pake, G. E., and Berson, R. "Structural Investigation by Means of Nuclear Magnetism. III. Ammonium Halides." *J. Chem. Phys.* 22, no. 4 (1954): 643–650.
80. Woessner, D. E. and Snowden, B. S., Jr. "Temperature Dependence Studies of Proton and Deuteron Spin-Lattice Relaxation in Ammonium Chloride." *J. Phys. Chem.* 71 (1967): 952–956.

81. Shimomura, K., Kodama, T., and Negita, H. "Proton Spin-Lattice Relaxation in NH_4Cl." *J. Phys. Soc. Jpn.* 31 (1971): 1291–1295.
82. Bashirov, F. I. "Proton Spin-Lattice Relaxation in Monocrystalline Ammonium Chloride." *J. Magn. Reson. A* 222 (1996): 1–8.
83. Bashirov, F. I. "Spontaneous Break of Symmetry in Molecular Crystals." *Crystallographia* 46, no. 3 (2001): 494–499. (Engl. Transl.: *Crystallogr. Rep.* 46, no. 3 (2001).
84. Woessner, D. E. and Snowden, B. S., Jr. "Proton Spin-Lattice Relaxation Temperature Dependence in Ammonium Bromide." *J. Chem. Phys.* 47 (1967): 378–381.
85. Woessner, D. E. and Snowden, B. S., Jr. "Spin-Lattice Relaxation and Phase Transitions in Deuterated Ammonium Bromide." *J. Chem. Phys.* 47, no. 7 (1967): 2361–2363.
86. O'Reilly, D. E. and Tsang, T. "Magnetic Resonance Studies of Ferroelectric Methylammonium Alum." *Phys. Rev.* 157, no. 2 (1967): 417–426.
87. Kydon, D. W., Petch, H. E., and Pintar, M. "Spin-Lattice Relaxation Times in Deuterated Ferroelectric Ammonium Sulphate and Fluoroberyllate." *Chem. Phys.* 51, no. 2 (1969): 487–491.
88. Brown, R. J. C. "Proton Magnetic Relaxation in NH_4IO_4." *J. Magn. Reson.* 42 (1981): 1–4.
89. Rigny, P. and Virlet, J. "NMR Study of Molecular Motions near the Solid-Solid Transition in the Metal Hexafluorides." *Chem. Phys.* 51, no. 9 (1969): 3807–3819.
90. Shenoy, R. K., Sundaram, C. S., and Ramakrishna, J. "Proton Spin-Lattice Relaxation in Sodium Ammonium Selenate Dehydrate $NaNH_4SeO_4 \times 2H_2O$." *Phys. Stat Sol. A* 62, no. 1 (1980): 93–96.
91. El Saffar, Z. M., Schultz, P., and Meyer, E. F. "Proton Magnetic Resonance Studies of Trimethylacetonitrile." *J. Chem. Phys.* 56, no. 4 (1972): 1477–480.
92. Albert, S., Gutowsky, H. S., and Ripmester, J. A. "NMR Relaxation Studies of Solid Hexamethylethane and Hexamethyldisilane." *J. Chem. Phys.* 56, no. 3 (1972): 1332–1336.
93. Albert, S., Gutowsky, H. S., and Ripmester, J. A. "On a T_1 and $T_{1\rho}$ Study of Molecular Motion and Phase Transitions in the Tetramethylammonium Halides." *J. Chem. Phys.* 56, no. 7 (1972): 3672–3676.
94. Stejscal, E. O., Woessner, D. E., and Farrar, T. C. "Proton Magnetic Resonance of the CH_3 Group. V. Temperature Dependence of T_1 in Several Molecular Crystals." *J. Chem. Phys.* 31 (1959): 55–68.
95. Anderson, J. E. and Slichter, W. A. "Nuclear Spin-Lattice Relaxation in Solid *n*-Alkanes." *J. Phys. Chem.* 65, no. 9 (1965): 3099–3105.
96. Allen, P. S. and Cowking, A. J. "Nuclear Magnetic Resonance Study of Molecular Motions in Hexamethylbenzene." *Chem. Phys.* 47, no. 11 (1967): 4286–4289.
97. Boud, M. F. and Hubbard, P. S. "Nonexponential Spin-Lattice Relaxation of Protons in Solid CH_3CN and Solid Solution of CH_3CN in CD_3CN." *Phys. Rev.* 170, no. 2 (1968): 384–390.
98. Zaripov, M. R. "Effect of Intramolecular Motions to Nuclear Magnetic Relaxation in Some Organic Polycrystals." PhD diss., KFTI AN USSR Kazan, 1969.
99. Albert, S. and Ripmeester, J. A. "NMR Static and Rotating Frame Relaxation Studies in Solid $Cl_3CC(CH_3)_2Cl$." *J. Chem. Phys.* 59, no. 3 (1973): 1069–1073.

100. Kumar, A. and Jonson, C. S., Jr. "Proton Spin-Lattice Relaxation Studies of Reorienting Methyl Groups in Solids." *J. Chem. Phys.* 60, no. 1 (1974): 137–146.
101. Pratt, J. C., Watton, A., and Petch, H. E. "Deuteron Spin-Lattice Relaxation from Hindered Rotations in Molecular Crystals." *J. Chem. Phys.* 73 (1980): 3542–3546.
102. Eguchi, T., Soda, G., and Chinara, H. "Molecular Motions in Polimorphic Forms of Ethanol as Studied by Nuclear Magnetic Resonance." *Mol. Phys.* 40, no. 3 (1980): 681–696.
103. Yamchi, J. and McDowell, C. A. "NMR Investigations of N,N,N',N'-Tetramethyl-p-Phenylenediamine and Its Oxidized Tetrafluoroborate Salts (TMPD)." *J. Chem. Phys.* 75, no. 3 (1981): 1060–1068.
104. Christoforov, A. V., Bashirov, F. I., and Yuldasheva, L. K. "Influence of Steric Structure of Molecule On Internal Rotations in 2,2,4,4,7,7-hexamethyl-2-sila-1,3-dioxepane." *J. Struct. Chem.* 26, no. 5 (1985): 180–181.
105. Eguchi, T. and Chinara, H. "^{1}H Spin-Lattice Relaxation in Solid Methyl Chloride." *J. Magn. Reson.* 76 (1988): 143–148.
106. Mallory, F. B., Mallory, C. W., Conn, K. G., and Beekman, P. A. "Methyl Reorientation in Methylphenanthrenes. II. Solid State Proton Spin-Lattice Relaxation in the 1-CH_3, 9-CH_3 and 1-CD_3, 9-CD_3 Systems." *J. Phys. Chem. Solids* 51 (1991): 129–134.
107. Ganesan, K., Damle, R., and Ramakrishna, J. "NMR Study of Molecular Dynamics in Hydrazine Phosphates." *J. Phys. Chem. Solids* 51 (1991): 297–301.
108. Idziak, S., Haeberlen, U., and Zimmerman, H. "The Configuration and the Dynamics of the CD_3 Groups in Methyl Deuterated Dimethylalonic Acid: A Single Crystal Deuteron NMR Study $C(CD_3)_2(COOD)_2$." *Mol. Phys.* 73, no. 3 (1991): 571–583.
109. Reynhardt, E. C., Jurga, S., and Jurga, K. "The Structure and Molecular Dynamics of Solid n-Decylammonium Chloride." *Mol. Phys.* 77, no. 2 (1992): 257–278.
110. O'Reilly, D. E., Peterson, E. M., and Lammert, S. R. "Proton Magnetic Resonance of Ammonia at Low Temperatures." *J. Chem. Phys.* 52, no. 4 (1970): 1700–1703.
111. Andreev, N. K. "Thermal Motion of Coordinated Atomic Groups and NMR-Relaxation in Solids." PhD diss., Kazan State University, Kazan, 1976.
112. Mehring, M. and Raber, H. "Nonexponential Spin-Lattice Relaxation and Its Orientation Dependence in a Three-Spin System." *J. Chem. Phys.* 59, no. 3 (1973): 1116–1120.
113. Parker, R. S. and Schmidt, V. H. "Dipolar Relaxation of ^{7}Li by Hindered Rotations ND_XH_{3-X} in Lithium Hidrazinium Sulphate." *J. Magn. Reson.* 6 (1972): 507–515.
114. Van Hove, L. "Correlations in Space and Time and Born Approximation Scattering in Systems of Interacting Particles." *Phys. Rev.* 95, no. 1 (1954): 249–262.
115. Sears, V. F. *Thermal-Neutron Scattering Lengths and Cross-Sections for Condensed Matter Research.* Chalk River, Ontario: Chalk River Nuclear Laboratories, 1984.
116. Bashirov, F. I. "Angular Auto-Correlation Functions in Molecular Crystals and Liquids: Application to Incoherent Neutron Scattering Law." *Mol. Phys.* 99 (2001): 25–32.

117. Bashirov, F. I. and Gaisin, N. K. "Function of Incoherent Scattering of Neutrons in Molecular Crystals and Liquids." *Khimicheskaya Fizika* 21, no. 3 (2002): 32–40.
118. Rowe, J. M., Livingston, R. C., and Rush, J. J. "Neutron Quasielastic Scattering Study of SH-Reorientation in Rubidiumhydrosulfide in the Intermediate Temperature Trigonal Phase." *J. Chem. Phys.* 59, no. 12 (1973): 6652–6655.
119. Rush, J. J., de Graaf, L. A., and Livingston, R. C. "Neutron Scattering Investigation of the Rotational Dynamics and Phase Transitions in Sodium and Cesium Hidrosulfides." *J. Chem. Phys.* 58, no. 8 (1973): 3439–3448.
120. Steenbergen, Chr., de Graaf, L. A., Bevaart, L., Dartolome, J., and de Jongh, L. J. "Rotational Motion of NH_4^+ Group in NH_4ZnF_3 Studied by Quasielastic Neutron Scattering." *J. Chem. Phys.* 70, no. 3 (1979): 1450–1455.

Index

F

G

H

K

L

M

N

T

Milton Keynes UK
Ingram Content Group UK Ltd.
UKHW040052071024
449327UK00019B/499